'Remarkable.'

TLS on *Weird Maths*

'A glorious trip through some of the wilder regions of the mathematical landscape, explaining why they are important and useful, but mostly revelling in the sheer joy of the unexpected. Highly recommended!'

Ian Stewart, author of *Significant Figures*, on *Weird Maths*

'Darling and Banerjee take us on a captivating ride through a vast landscape of mathematics, touching on mesmerising topics that include randomness, higher dimensions, alien music, chess, chaos, prime numbers, cicadas, infinity, and more. Read this book and soar.'

Clifford A. Pickover, author of *The Math Book*, on *Weird Maths*

'In this inspired collaboration, a young maths prodigy teams up with a popular science writer to present a fresh view of the world of mathematics. Together they fearlessly tackle some of the most weird and wonderful topics in mathematics today.'

John Stillwell, Professor of Mathematics, University of San Francisco, and author of *Elements of Mathematics*, on *Weird Maths*

'A grand tour of the most exotic locations in the mathematical cosmos. *Weirder Maths* is exhilarating and entertaining, and will leave you with a wide-eyed appreciation of the world of numbers.'

Michael Brooks, author of *The Quantum Astrologer's Handbook*, on *Weirder Maths*

ALSO BY DAVID DARLING
AND AGNIJO BANERJEE

Weird Maths
Weirder Maths
Weirdest Maths

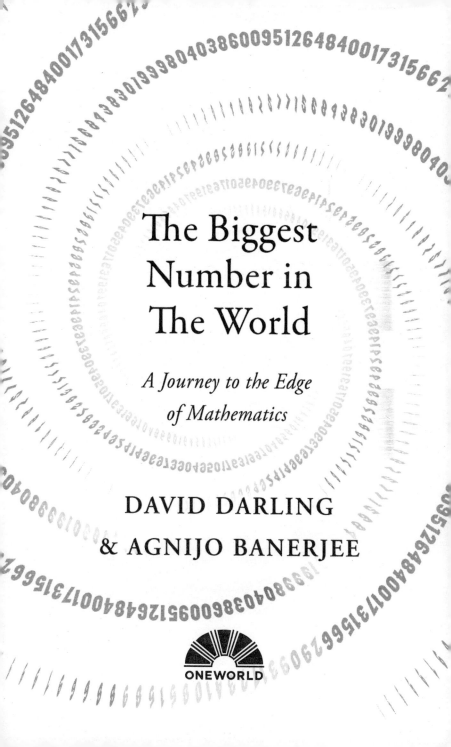

The Biggest Number in The World

A Journey to the Edge of Mathematics

DAVID DARLING
& AGNIJO BANERJEE

ONEWORLD

A Oneworld Book

First published by Oneworld Publications in 2022

Copyright © David Darling and Agnijo Banerjee 2022

The moral right of David Darling and Agnijo Banerjee to be
identified as the Authors of this work has been asserted by them in
accordance with the Copyright, Designs, and Patents Act 1988

ISBN 978-0-86154-305-2
eISBN 978-0-86154-306-9

Illustration credits: Sand dunes in the Sahara © Luca Galuzzi (www.galuzzi.it);
German Stamps courtesy of the National Liberation Museum 1944–1945 (Wikimedia
Commons); Sierpiński triangle © Beojan Stanislaus (Wikimedia Commons); Ronald
Graham portrait © Cheryl Graham (Wikimedia Commons); An artist's impression
of a Pioneer spacecraft on its way to interstellar space courtesy of NASA/Don
Davis (Wikimedia Commons); John H. Conway © Thane Plambeck (Wikimedia
Commons); Model of a Turing machine © GabrielF (Wikimedia Commons).

Typeset by Tetragon, London
Printed and bound in Great Britain by Clays Ltd, Elcograf S.p.A.

Oneworld Publications
10 Bloomsbury Street
London WC1B 3SR
England

Stay up to date with the latest books,
special offers, and exclusive content from
Oneworld with our newsletter

Sign up on our website
oneworld-publications.com

MIX
Paper from
responsible sources
FSC® C018072

The study of mathematics, like the Nile, begins in minuteness but ends in magnificence.

—Charles Caleb Colton

The more you approach infinity, the deeper you penetrate terror.

—Gustave Flaubert

Contents

Introduction

THE PHYSICAL UNIVERSE is vast beyond imagining. Even the nearest star lies at a distance almost impossible to grasp with our Earth-bound brains. The edge of the observable universe is inconceivably more remote: about 46 billion light-years, or 270 billion trillion miles, away. Yet we're about to embark on a much greater voyage, not into the depths of space but into the farthest reaches of the mathematical cosmos.

Along the way we'll come across some extraordinary ideas, so alien to our normal way of thinking that the biggest challenge will be to find familiar words and concepts by which we can build bridges to understanding. We'll venture far from home into regions of thought that, until now, few have seen or experienced. Our quest: nothing less than to find the edge of the numerical universe.

Surely, you might say, there *is* no such edge. Numbers go on forever. Even if we were to fill this book, or a library full of books, with 1 followed by zeros – or all 9s – on every line on every page, at the end you could name a bigger number simply by saying 'and 1'. And that's true. The number line stretches away into the mists of infinite distance. But, as we're about to find out, the search for an ultimately large number isn't confined to trekking slowly, step by step, down an endless road. There are some surprising, mind-bending alternatives to the often-repeated mantra 'there's no biggest number'. Some of these will involve entering a shadowy land, still largely unexplored, between the finite and the infinite. Others will transport us into what are effectively parallel universes of maths, where different rules operate and what we thought was secure knowledge is easily overturned.

As with any expedition into the unknown we need to go well prepared. We'll look at the history of large numbers and how the subject has been mapped out up to this point. We'll delve into a few areas – fascinating in themselves – that are rarely broached in school or university curricula in order to equip ourselves for the great quest ahead.

Like mountaineers attempting to climb previously unconquered peaks, certain mathematicians throughout history have had the courage to try to scale new heights in the mighty ranges of towering numbers. Often they've ventured alone, not relying on the intellectual, moral or financial support of others to help them in their ambition. These pioneers of a strange land have had to develop new tools and techniques to go beyond what was possible before. And the vistas with which they've been rewarded are no less breathtaking and spectacular, in their way, than the views from the summits of Everest or the Matterhorn. These are the mind's-eye spectacles that await us in the pages ahead.

We also have personal reasons for writing this book. Number theory – and the mathematics of very large numbers, in particular – is a passion of Agnijo's. It's a subject that's fascinated him throughout his school career, which culminated in him taking first place in the 2018 International Mathematical Olympiad, and as a student at Cambridge. David has always enjoyed finding ways to explain difficult ideas to a wide audience. The book is the culmination of an unusual writing partnership, which began while David was tutoring Agnijo as a young teenager.

There's a widespread suspicion that maths is cold and austere, somewhat aloof from the real world of people. But nothing could be further from the truth. Mathematics, along with music and art, is among the most human of enterprises, steeped in passion, tragedy, comedy and romance, wild and wonderful characters, and bold new ideas that constantly threaten the status quo. Nowhere is this drama of maths more evident than in the ultimate intellectual challenge: the search for the biggest number in the world.

CHAPTER 1

Of Sand and Stars

ARE THERE MORE grains of sand on Earth or stars in the universe? With your eyes alone you can see at least a couple of thousand stars on a clear night well away from artificial lights, and nearer 4,000 if it's moonless and your eyesight is keen. In a handful of sand are many more grains than that. But space is huge, dauntingly so, and powerful telescopes reveal that it contains a host of galaxies, each harbouring billions of stars. On the other hand, the deserts, beaches and ocean beds of our planet are home to sand particles in dizzying profusion. So, sand or stars, which wins in the numbers game?

A study carried out by researchers at the University of Hawaii in 2003 estimated the number of sand grains in the world to be 7.5 million trillion, or 75 followed by 17 zeros. As for stars, the figure they came up with, for the whole of the observable universe, was 70 thousand million trillion. That's about ten thousand stars for every sand grain.

The Greek mathematician and scientist Archimedes was also interested in this kind of problem. In the third century BCE he wrote a short treatise, addressed to Gelon, King of Syracuse, that's come to be known as 'The Sand Reckoner'.

FIGURE 1.1: Sand dunes in the Sahara, Libya.

Sometimes described as the first research-expository paper, because it combines both accuracy and clear language, aimed at the layperson, it asks: How many grains of sand would fit in the universe?

The answer, of course, depends on how big is an average grain of sand and how big is the universe. Archimedes figured, very generously (to the point of being unrealistic), that one poppy seed could contain 10,000 grains of sand, which would make the grains almost microscopic in size. He also reckoned that 40 poppy seeds, side by side, would stretch across one Greek *dactyl*, or finger-width, equal to about three quarters of an inch (19 millimetres). A sphere one dactyl wide would then be able to hold in the region of 640 million sand grains.

As for the size of the universe, Archimedes based his estimate on the classical heliocentric theory of his predecessor

Aristarchus. In this model of space, Earth orbits around the Sun while the stars are fixed to a sphere, also centred on the Sun, but much further out. The fact that the Greeks couldn't discern any change in the relative positions of stars in the sky – a so-called parallax – as Earth moved from one side of the Sun to the other meant that stars had to lie a certain minimum distance away. This gave Archimedes his estimate for the smallest possible diameter of the then-known universe – in modern units, about two light-years.

Today we can easily do the maths and arrive at how many Archimedean-sized sand grains would fit inside a ball two light-years wide. The answer comes out to be roughly one followed by 63 zeros, which can be written compactly as 10^{63} – meaning $10 \times 10 \times 10 \times \ldots \times 10$ (with 63 tens). The problem Archimedes faced is that our handy ways of representing big numbers didn't exist in his day. The Arabic numerals, 0 to 9, that we now use, emerged about 800 years later (and in India, not Arabia). Place-value notation, in which the same symbol is used to represent different orders of magnitude depending on its position (for example, the '3' in 30, 300, and 3,000) was still in its infancy in Babylon but hadn't yet reached Greece. And there was in those days no such thing as index notation, in which how many times a number must be multiplied by itself is written as a superscript (as in 10^{63}).

At the time when Archimedes began his cosmic sand calculations the Greeks used letters of the alphabet to represent numerals. A different letter stood for the equivalent of our numbers 1 to 9, multiples of ten from 10 to 90, and multiples of a hundred from 100 to 900. The familiar 24 letters, alpha to omega, which have survived in present-day Greek, had to be supplemented by others taken from older languages and dialects to provide enough labels. Alpha to theta stood for 1 to 9, iota to koppa (borrowed from the Phoenician) for multiples of ten from 10 to 90, and rho to sampi (used in some eastern Ionic dialects) for multiples of a hundred from 100 to 900. The Greeks didn't use the same letter again and again in different positions, so that, for example, 222 would be written as σκβ (sigma kappa beta = 200 + 20 + 2). For multiples of a thousand, from 1,000 to 9,000, some of the same letters were employed but with various extra marks. And that was as far as the ancient Greek labelling system of numerals went, except for the *murious* – the largest single unit defined, written as a capital mu (M) and equivalent to our 10,000. The Romans called it the *myriad*, a name that became absorbed into English but with the altered meaning of 'countless' or a very large (but undefined) number.

The Greeks could write numbers that were bigger than a *murious* but only as multiples of M using strings of letters in the manner described. For example, 1,234,567 would be written as ρκγM ,δφξζ (123 × 10,000 + 4,567). It's an approach that quickly runs out of steam for anything beyond what we would call a few hundred million.

Archimedes realised that to represent the kind of gigantic numbers that would arise from his cosmic sand calculations, he'd have to come up with a whole new system of number naming. He started by defining anything up to a myriad myriad as being a number of the 'first order'. To us that mightn't seem like a big step because we can easily write a myriad myriad as $10^4 \times 10^4$, which equals 10^8 (a hundred million), and then carry on indefinitely from there. But there was nothing like our index notation, in which an index or exponent is used to show how many times a number must be multiplied by itself, when Archimedes took on his big-number project.

Having defined any number up to a myriad myriad as belonging to the first order, he moved on to numbers that lay between a myriad myriad and a myriad myriad times a myriad myriad (1 followed by 16 zeros, or 10^{16} in modern notation). These, he said, belonged to the 'second order'. Then he progressed to the third order, and the fourth, and so on, in the same way – each successive order being a myriad myriad times larger than the numbers of the previous order. Eventually, he reached numbers of the myriad myriadth order, in other words, in our index notation, 10^8 multiplied by itself 10^8 times, or 10^8 raised to the power 10^8, which equals $10^{800,000,000}$. All these numbers, of which the largest would have 800 million digits if written out in full, he defined as belonging to the 'first period'. The number $10^{800,000,000}$ itself he took to be the springboard for the second period, at which point he began the process all over again. He defined orders of the second period by the same method, each new order being a myriad myriad times greater than the last, until, at the end of the myriad myriadth period, he'd reached the colossal value of a myriad myriad raised to the power of a myriad myriad times a myriad myriad, which we'd write as $10^{80,000,000,000,000,000}$, or 10 to the power of 80 thousand trillion.

Remember, Archimedes had no knowledge of our compact ways of writing big numbers. There wasn't even the concept of zero in ancient Greek maths. Starting from a system that struggled to name numbers that were bigger than a few hundred million, he fashioned a method to describe a number that, in decimal form, would have 80 thousand trillion digits.

For the purposes of his sand-counting project, it turns out, Archimedes didn't need numbers anywhere near this large. Using his estimates of the size of a grain of sand and of the whole universe, he came up with a value that was only of the eighth order of the first period. In index notation, a mere 8×10^{63} or so of Archimedes' minuscule grains would have been enough to pack the two-light-year-wide Greek cosmos full of sand. Even using a modern, and much larger, estimate for the diameter of the observable universe of 92 billion light-years there'd be no room for more than about 10^{95} sand grains – still a number of just the twelfth order, first period.

'The Sand Reckoner' was cutting-edge stuff. Not only did Archimedes offer a picture of the universe that most closely resembles what we know today, given the limited data he had available, but he invented a whole new way of describing big numbers. He was the first person to tackle the problem of naming and manipulating large numbers without the benefit of modern notation. Using a system with base 10,000, he effectively pioneered exponentiation – the process of raising one quantity to the power of another. He also discovered the law of adding exponents, namely $x^m \times x^n = x^{m+n}$, for any numbers x, m, and n; for example, $3^2 \times 3^3 = (3 \times 3) \times (3 \times 3 \times 3) = 3^5$.

Archimedes was the first person to show that it's possible to go beyond the tradition of his era of simply calling huge numbers of things 'innumerable'. Sand and stars, in particular, came in a lot for this kind of treatment. The Greek poet Pindar, who predated Archimedes, wrote in his *Olympus*

Ode II: 'sand escapes counting'. There's even a Greek word, *psammakósioi* – literally, 'sand-hundred' – that's used to mean 'uncountable'. Writers of the Bible, too, gave up on the arithmetic of sand and stars. The phrase in Genesis (32:12) 'the sand of sea, which cannot be counted for multitude' is one of twenty-one Biblical references suggesting that it's impossible to put a figure on the numbers of sand grains out there. Hebrews (11:12) conflates the two: 'So many as the stars of the sky in multitude, and as the sand which by the seashore is innumerable.'

As we've seen, Archimedes didn't confine himself to the sand on a seashore or even on the Earth as a whole. He made sure that none of his contemporaries could possibly outdo his number count by imagining the entire universe to be packed full of sand grains so small that they'd be barely visible. It would be interesting to know what he'd have thought of the efforts of other intellectuals, a few hundred years later, who also wrote about large numbers but in a different part of the world, and for a very different reason.

Eastern philosophy, and Buddhism in particular, has always been fascinated with the vastness of space, time, and mind. It's not surprising, then, that scholars of these thought systems eventually came around to putting numbers to the age or extent of things on the broadest of cosmic scales. In one of the major scriptures, or sutras, of Mahayana Buddhism, written in the third century CE and known as *Lalitavistara* (Sanskrit for 'The Play in Full'), a conversation takes place between Gautama Buddha, who'd died hundreds of years earlier, and a mythical mathematician named Arjuna. In reply to a question by Arjuna, the Buddha launches into a head-spinning exposition of a system of numerals based on the *koti*, a Sanskrit term for ten million (10,000,000). At each step the Buddha names a number that's one hundred times greater than the last: one *ayuta* is 100 *koti*, one *niyuta* is 100 *ayuta*, and so on, until he reaches the *tallakshana*, which equals one followed by 53 zeros. Beyond the range of the *tallakshana*, the Buddha explains, lies another, that of the *dvajagravati*, which takes us to 10^{99}, and then four others in an ascending hierarchy that reaches up to the *uttaraparamanurajahpravesa*, equivalent to 10^{421}.

Impressive though this number is, the Buddha's only just getting into his stride. In the *Avatamsaka* ('Flower Garland' sutra) he reveals a different and hugely more powerful counting system. In Thomas Cleary's translation of chapter 30 of the *Avatamsaka*, the Buddha explains how the system starts off:

> Ten to the tenth power times ten to the tenth power equals ten to the twentieth power; ten to the twentieth power times ten to the twentieth power is ten to the fortieth power...

Then he continues, in exasperating detail, squaring each successive number, to yield 10^{80}, 10^{160}, 10^{320}, and so on, until, after a couple more scrolls-worth of itemisation, he arrives at $10^{101,493,392,610,318,652,755,325,638,410,240}$. For some reason, unfortunately not explained in the sutra, the Buddha considers this number to mark some kind of limit. Further squaring, he says, leads to a number called 'incalculable'. Next he moves to squaring the square – in other words, raising to the fourth power. 'Incalculable' to the fourth power gives 'measureless';

repeating the process leads to 'boundless'. After some more similar steps, and excursions into the Sanskrit thesaurus, we're led to 'unspeakable', the raising to the fourth power of which culminates in 'untold'. Then, in a final flourish, the Buddha reports that

> Untold unspeakables
> Fill all unspeakables
> In unspeakable aeons
> Explanation of the unspeakable cannot be unfinished

Quite why those who wrote the *Avatamsaka* had the Buddha stop doing precise maths at 'incalculable' and resort instead to a string of superlatives isn't clear. Maybe they got bored with writing out long lists of numerals or perhaps scrolls were in short supply. The likeliest reason, though, is that they wanted to give the impression that, in the end, the universe extends beyond ordinary logic and analysis into a realm accessible only to those who are enlightened.

In any event, we can easily take the mystery out of all these shenanigans. The mighty untold, far from being incalculable or untellable, works out in reality to be, in modern notation

$$10^{10 \times 2^{122}}$$

or approximately $10^{53,000,000,000,000,000,000,000,000,000,000,000,000}$. It's obviously a truly vast number. Archimedes would doubtless have been impressed with it because it dwarfs the biggest number he reached in 'The Sand Reckoner'. Archimedes maxed out at $10^{80,000,000,000,000,000}$, whereas to reach 'untold' you'd have to multiply Archimedes' number by itself roughly 660 million trillion times.

Both Archimedes and the Buddhist sutras used large numbers to give some impression of the immensity of their respective versions of the universe. With Archimedes it was more a scientific enterprise, whereas the Eastern goal seems to have been to inspire reverence for a holistic vision of the cosmos, inaccessible to conventional thought. These were early and isolated peaks in the quest to describe ever-larger numbers. Only comparatively recently, within the past century and a half, have mathematicians given much thought to looking at what lies beyond these seminal insights: at numbers that are incomparably larger and, as a consequence, demand innovations in order that they can be represented in a manageable way.

For most practical purposes, whether it's everyday conversation, economics, or measurement in science, we use words ending in '-illion' to name big numbers. The current world population is about 7.8 billion, the nearest star (Proxima Centauri) lies at a distance of 40.2 trillion kilometres, and so on. It's a method of number naming that has its roots in late mediaeval times when the word 'million' began to appear in writings such as Chaucer's *The Canterbury Tales*. 'Million' comes from the Italian *millione*, which, in turn, stems from the Latin *mille* for 'thousand' and the augmentative suffix *-one* (hence 'million' = 'thousand thousand'). 'Bymillion' (a million million) and 'trymillion' (a million million million) were in circulation by the 1470s and, in 1484, the Frenchman Nicolas Chuquet proposed a complete systematisation of number names using words ending in -illion (or -yillion).

Not much is known about Chuquet beyond that he was born in Paris, held a bachelor's degree in medicine, and later moved to Lyon, where he died in his thirties. He certainly wasn't an eminent mathematician. He's remembered today for only one achievement: an article called *Triparty en la science des nombres* (The Science of Numbers in Three Parts), which wasn't published until 1880, almost four centuries after his death, by the linguist Aristide Marre who discovered Chuquet's manuscript. It then became clear that a student of Chuquet's, Estienne de la Roche, had essentially plagiarised his teacher's writings in the first part of his handbook of algebra, *l'Arismetique* (1520).

In his original work, Chuquet wrote down a very large number – 745324804300070002365321 – and then used marks to break it down into groups of six digits, starting from the right. Up to the first mark were the millions. Thereafter:

> *The second mark byllion, the third mark tryllion, the fourth quadrillion, the fifth quyillion, the sixth sixlion, the seventh septyllion, the eighth ottyllion, the ninth nonyllion and so on with others as far as you wish to go.*

We still use these names, with the 'y' replaced by 'i' and other minor changes, today. The only difference is that it's become widely accepted in English-speaking countries and some others that a billion is a thousand million (10^9) rather than a million million (10^{12}). Two different naming systems for large numbers emerged, which, in 1974, French mathematician Geneviève Guitel described as the 'long scale' and the 'short scale'. In the former, each term after a million – a billion, a trillion, and so on – is defined to be a million times larger than the one before, whereas in the latter the jumps are by a factor of a thousand. In British English both systems were used up until the mid-1970s or so. Today, the short scale, long favoured in North America, has been adopted in most of the English-speaking and Arabic-speaking world, as well as in Brazil, Russia, and several other countries, while the long scale remains in vogue elsewhere. The system can easily be extended beyond a trillion, using the prefixes quad-, quin-, sex-, sept-, and so on. A quadrillion, for instance, using the short scale, is a thousand times greater than a trillion, or 10^{15}; a quintillion is a thousand times greater than a quadrillion, or 10^{18}, and so on. Each multiple of a thousand bumps the prefix up by one. A centillion ('cent-' meaning a hundred) is the same as 1 followed by 303 zeros and is the biggest number listed in standard dictionaries with a name that uses this convention: that every additional three zeros advances by one the Latin or Greek number prefix.

Until a few centuries ago, there was really no practical need to have names for numbers that were much bigger than a million – unless you were doing something unusual like counting sand grains or extolling Eastern philosophy. It wasn't until the start of the nineteenth century that the world's population ticked past a billion, and later still that atoms were discovered and astronomers started to appreciate how many stars were in our galaxy, never mind all those that lay beyond. But pure mathematicians aren't confined by the limits of physical reality and, early on, they realised that numbers went on and on, eventually far exceeding any system that had been devised for their description. By the dawn of the Renaissance, it had become simply unacceptable that numbers could exist for which there were neither convenient names nor ways to represent them.

Chuquet systematised the '-illion' way of naming numbers, while Archimedes and others, including Muhammad ibn Musa al-Khwarizmi, in ninth-century Persia, and Abu'l Hasan ibn Ali al Qalasadi, in the mid-fifteenth century, laid the basis for exponentiation. The word 'exponent' itself was coined in 1544 by German mathematician and monk Michael Stifel. Finally, in the early sixteenth century, French mathematician and philosopher René Descartes, in Book I of his text *La Géométrie*, introduced notation of the form x^n or 'x to the n' (although, at the time, he was thinking more in terms of geometry than algebra). In the expression x^n, x is a number, known as the base, and n is the index or exponent. It's also common for n to be called the 'power' of x, although strictly speaking if $a = x^n$ then a is the power, not n.

Just as multiplication can be thought of as repeated addition ($4 \times 3 = 3 + 3 + 3 + 3$), exponentiation is a compact way of writing and performing repeated multiplication ($6^5 = 6 \times 6 \times 6 \times 6 \times 6$). For almost every purpose we need – except many of those we'll meet in this book! – exponentiation, or working in index form, is sufficient for dealing with even extraordinarily large numbers. A number such as 100,000,000,000,000,000,000 – a hundred million trillion – can be written compactly as 10^{20} and described as '10 to the power 20' or just '10 to the 20'.

For the most part, describing big numbers in terms of '-illions' works just fine. But sometimes it's good to have a special name for a particular large number. One day in 1920, American mathematician Edward Kasner was walking with his nephews, nine-year-old Milton Sirotta and his brother Edwin, by the Palisades (the cliffs that line the Hudson River in New Jersey). Kasner got talking to them about numbers and how big they could be – as big as, say, one followed by a hundred zeros. Writing later in *Mathematics and the Imagination* (1940), which he co-authored with James Newman, Kasner recalled: '[Milton] was very certain that this number was not infinite, and therefore equally certain that it had to have a name.' The name he came up with was 'googol'. At the same time, young Milton suggested 'googolplex' for a number that was even bigger. Kasner wrote:

> A googolplex is much larger than a googol, but is still finite, as the inventor of the name was quick to point out. It was suggested that a googolplex should be 1 followed by writing zeros until you get tired. This is a description of what would happen if one actually tried to write a googolplex, but different people get tired at different times and it would never do to have Carnera [a heavyweight boxing champion] a better mathematician than Dr. Einstein, simply because he had more endurance.

Kasner offered a more precise definition of a googolplex as '1 followed by a googol number of zeros', or 10^{googol}. Whereas a googol, albeit hard to imagine, is easy to write out in full:

10,000,000,000,000,000,000,000,000,000,000,
000,000,000,000,000,000,000,000,000,000,000,
000,000,000,000,000,000,000,000,000,000

a googolplex is sensationally larger. There isn't enough paper on Earth, or, come to that, matter in the entire observable universe, to write out the digits of a googolplex, not even if you wrote each zero as small as a subatomic particle. The googolplex utterly dwarfs any number named in antiquity, including the mighty 'untold'.

Mention 'googol' or 'googolplex' and most people will instantly think of the ubiquitous search engine or the place where it's now headquartered. In 1996, the founders of what would become Google, Stanford PhD students Larry Page and Sergey Brin, were working out of a makeshift office – a garage rented to them by mutual friend and future YouTube CEO Susan Wojcicki – in Menlo Park, California. They'd called their prototype search engine 'BackRub' because it analysed the Web's back links (incoming links to a webpage). But as their search technology rapidly improved they sought a more commercially appealing name for the new product. In September 1997, Page and his office mates held a brainstorming session, complete with whiteboard, in 'Susan's garage' to think of something that would work – a word that captured the idea of indexing a huge amount of data. One of those present, graduate student Sean Anderson, suggested 'googolplex', which Page immediately shortened out loud to 'googol'. Anderson, sat at his computer terminal, checked the internet domain registry to see if the name was still available for registration and use. But making the mistake of thinking that the word was spelled 'google' he checked 'google.com' instead of 'googol.com', and found it to be available. Page liked the name and within hours had registered 'google. com' on behalf of himself and Brin.

The name is certainly suggestive of the immense volume of data now involved in web indexing. In 2017 Google reported that it stored information on about 30 trillion pages. Google, Microsoft, Amazon, and Facebook between them hold at least 1,200 petabytes (1.2×10^{15} bytes) of data – a figure that's rising fast, month by month. If Google were to maintain its average annual rate of growth in indexing over the next few centuries (unlikely!), it would have indexed a googol pages by the year 2698.

In ancient times only a few intellectuals, such as Archimedes, glimpsed how very large numbers might be relevant to the real world. But today we're all familiar with hearing about billions and trillions of things, and scientists and mathematicians find uses for numbers that make even the googol seem small. Can we truly grasp the size of these numbers, let alone the vastness of others we'll encounter later on in our quest to find the biggest number of all? No, not even the greatest mathematical genius can do that. But what we can do is try to find words or concepts that form a bridge between the world with which we're familiar – the world we can sense or construct in our imagination – to numbers that lie far beyond the capacity of the physical universe to contain.

CHAPTER 2

At the Limits of Reality

NUMBERS BOTH HUGE and tiny abound in science – and for obvious reasons. The universe is incredibly large and the particles of which it's ultimately composed are fantastically small. We can quickly end up going well past a trillion if we start to count the numbers of very small things in nature (such as atoms) or measure on a cosmic scale with units that seem reasonable for everyday purposes. For instance, the metre is a sensible unit in the human world but not so much when we start talking about interstellar distances. Even the nearest star to the Sun, Proxima Centauri, lies about 40,000,000,000,000,000 – 40 thousand trillion – metres away.

The routine appearance of very large and very small numbers in science is why 'scientific notation' is commonly used. In scientific notation, also known as standard form or index form, 40 thousand trillion is written compactly as 4×10^{16}. This makes it easy to see at a glance how many zeros come after the 4.

Another way to make the numbers we're dealing with more manageable is to use bigger units. That's why astronomers often talk about distances in terms of light-years or parsecs. One light-year is the distance travelled by light, moving at a speed of 300 million metres per second, in one year. It equals about 9.46 trillion kilometres, so that the distance to Proxima comes out to be 4.24 light-years.

A parsec is the distance at which the average diameter of Earth's orbit around the Sun supports an angle of 1/3600th of a degree, known as one arcsecond. It works out to be 3.26 light-years. Proxima is 1.30 parsecs away and the centre of the Milky Way Galaxy a little over 8,000 parsecs from Earth. Even the parsec, though, starts to seem small once we move beyond our own galaxy and deep into the intergalactic void. Then astronomers turn to the kiloparsec, the megaparsec, and finally the gigaparsec – a billion parsecs. The distance across the entire observable universe is about 28.5 gigaparsecs or 8.8×10^{23} kilometres.

As we saw in the first chapter, numbers of this magnitude are nothing new in science. Archimedes' cosmic count of sand grains was around 8×10^{63} in modern notation. Fortunately for us, the universe isn't actually packed full of sand. Nevertheless, there are some other extremely big numbers in science that apply to real situations or, at least, our attempts to estimate them.

Back in 1811, Italian scientist Amedeo Avogadro proposed that the volume of a gas, at a given temperature and pressure, is proportional to the number of molecules within it regardless of the actual gas involved. This would mean that equal volumes of *different* gases, for example oxygen and carbon dioxide, under the same conditions, contained the same number of molecules. Although Avogadro believed in the existence of atoms and molecules, and drew a distinction between them, he'd no way of knowing their size. The first reasonably accurate measurements of what became known as Avogadro's constant were made in the early 1900s from experiments carried out by French physicist Jean Perrin. Today the value of Avogadro's constant is known accurately to be $6.02214076 \times 10^{23}$. It's defined to be the number of constituent particles, which may be molecules or atoms (or even ions), in a quantity of a substance called a mole. One mole is the molecular weight of a substance in grams. So, for instance, 31.9988 grams of oxygen and 44.0095 grams of carbon dioxide, under the same conditions, both contain $6.02214076 \times 10^{23}$ molecules.

By everyday standards, this number – six hundred billion trillion – is immense. In fact, it's the largest physical constant with which scientists deal on a routine basis. It also gives a feel for the minuscule nature of atoms and molecules. One mole of water weighs just 18 grams and occupies only a few drops, yet it contains six hundred billion trillion water molecules!

A much bigger number was named after another famous scientist, Sir Arthur Eddington. In his book *The Mathematical Theory of Relativity*, published in 1923, Eddington wrote:

> I believe there are 15, 747, 724, 136, 275, 002, 577,
> 605, 653, 961, 181, 555, 468, 044, 717, 914, 527, 116,
> 709, 366, 231, 425, 076, 185, 631, 031, 296 protons
> in the universe and the same number of electrons.

What's outrageous about this isn't the size of the number – after all, the number of protons and electrons across the whole universe is bound to be humongous – but the extra-ordinary precision with which it was stated. Had Eddington simply declared the total number of protons to be '*about* 1.57×10^{79}', or '*about* 15.7 million trillion trillion trillion trillion trillion', it wouldn't have created much of a stir. However, he claimed to have figured out the value down to the last particle!

By the time Eddington became interested in large cosmic numbers, he was already a world-renowned astrophysicist. In 1919, he led an expedition to observe a total solar eclipse in South Africa which confirmed one of the key predictions of Einstein's general theory of relativity – that the path of light from a star will be bent when it passes near a massive object (in this case, the Sun). He was also a pioneer investigator of stellar physics and the first to propose, in 1920, that stars generate their heat and light by the process of nuclear fusion.

During the 1920s, Eddington began to obsess more and more about building a grand theory that would unite relativity, quantum mechanics, cosmology and gravitation. Although it started out conventionally enough, his work soon began to take on board elements of numerology and aesthetics. He wasn't alone in this almost mystical fascination with what became known eventually as the 'large numbers hypothesis'. In 1919, German mathematician Hermann Weyl started the ball rolling by noting that the ratios of some basic distances and forces in nature were both very large and very similar. For instance, the electric force between a proton and electron is about 10^{40} times as big as the gravitational force between them. This same factor of 10^{40} cropped up when Weyl divided the radius of the universe, as it was then estimated to be, by what's called the classical electron radius.

As Eddington delved into these same kinds of relationships that link the submicroscopic world with the macroscopic, he became especially intrigued by an enigmatic factor in nature known as the fine-structure constant. This constant crops up in all kinds of places in atomic and nuclear physics. One of the things it does is to calibrate the strength of the electromagnetic force between elementary charged particles, such as electrons. Other physicists, right up to the present, have been struck by its pivotal role in cosmic affairs at different scales. Wolfgang Pauli had a lifelong fascination with the number, once commenting: 'When I die my first question to the Devil will be: What is the meaning of the fine-structure constant?' Richard Feynman referred to it bluntly as 'one of the greatest damn mysteries of physics'.

At the time Eddington first turned his attention to the fine-structure constant, its value wasn't known from experiment to any great accuracy. It was thought to be about 1/136. In a series of convoluted steps, Eddington claimed to have proved theoretically that the value was *exactly* 1/136 and, because of this, his reasoning led him to believe, the number of protons in the universe was 136×2^{256}. This is the infamous Eddington number, which he wrote out in full in his 1923 book and repeated again in a public lecture he gave in 1938 at Trinity College, Cambridge.

Unfortunately, later experiments led to the value of the fine-structure constant being adjusted downwards, putting it closer to 1/137. (In fact, it's now known to be 1/137.03599084.) That experimental readjustment didn't faze Eddington: he simply tweaked his theory so that it produced 1/137 exactly! But, not surprisingly, no one else was persuaded by such a convenient fudging of the issue. Other scientists lost faith in his large-number reasoning and the satirical magazine *Punch* captured the mood by dubbing him 'Sir Arthur Adding-One'. Eddington's number turned out to be a fiction. But it does retain one distinction: it's the biggest specific number – not an estimate or approximation – that's ever been claimed to have a bearing on the physical world.

As mentioned, when lots of little bits make up relatively gigantic things, big numbers are bound to crop up. The cells that comprise our bodies are microscopic in size and a few tens of trillions of them are needed to make up an average human being. What you're using to assimilate these ideas at the moment is a brain that contains about 86 billion neurons, or nerve cells. That's the number arrived at by a team of researchers at the University of Rio de Janeiro in 2009. Because each neuron is linked to many other neurons, the total number of connections in the brain is much larger than 86 billion: a figure of 100 trillion is in the right ballpark.

Every second, about 750,000 gallons of water flow over Niagara Falls, the great majority of it at the Horseshoe Falls. That amounts to 2.7 billion gallons per hour, 23.7 trillion gallons per year, or 1.9×10^{15} gallons over an average human lifetime. Given that there are about 1.27×10^{26} molecules in a gallon of water, this means that in the lifetime of a person who lives 80 years, roughly 2.4×10^{41} water molecules will have dropped over the Falls. At its current rate of flow, it would take about 13.8 million years for all the water on Earth to go over Niagara.

A useful way to grasp these kinds of numbers – millions, billions, trillions, and so on – is to think of cubes of dots.

Start off with a small number like a hundred, which might be how many people are at a modest-sized wedding reception. Imagine a cube that has a hundred dots on each edge. Then it will have 100 × 100 dots per face, and 100 × 100 × 100, or one million, dots altogether throughout its volume. Crowds of a million people sometimes come together, for instance at religious gatherings such as the hajj, the annual Islamic pilgrimage to Mecca. Somewhere between half a million and one million descended on the Woodstock festival in 1969. So we can grasp the size of a million, to some extent, through pictures and in our mind's eye. Moving on, a cube that has 1,000 dots on each edge will contain a total of a billion dots. A cube that has 100,000 dots along an edge – one for each person in a big, packed stadium – contains $10^5 \times 10^5 \times 10^5 = 10^{15}$, or one thousand trillion, dots, which is roughly as many as the number of connections between neurons in ten human brains.

To celebrate his first billion seconds of life (reached on 1 March 2020, during his thirty-first year), Dutch design engineer Daniel de Bruin invented a machine that takes more than a googol number of years to complete one rotation. Plenty of devices, from cars to washing machines, use reduction drives – a series of connected gears – to decrease the speed and increase the torque produced by a motor. De Bruin simply took this idea to an extreme by connecting 100 gears, each with a reduction ratio of 10 to 1, so that for every ten rotations of one gear, the next turns only once. The final gear in the chain rotates a googol – 10^{100} – times more slowly than the first. Since the first takes about four seconds to complete one revolution, the 100th gear has a rotation period of about 4×10^{100}

seconds. This is roughly 1.3×10^{93} years or about 100 billion trillion trillion trillion trillion trillion trillion times the current age of the universe!

De Bruin has already shared, on the internet, videos of his machine running for a full hour and has considered setting up a live stream for those who enjoy their action slow and predictable. Just don't expect much in the way of movement from the majority of the gear wheels – their motion would be undetectable even to the most sensitive instruments on Earth. As for the total energy required to turn the last gear around once, it would be far greater than all the energy in the universe combined.

Science isn't the only aspect of life that can spawn gigantic numbers. Sometimes, when a country's currency is hit by hyperinflation, almost everyone in the nation becomes a multi-trillionaire, virtually overnight, and yet may still be more or less penniless. It happened in the Weimar Republic in the early 1920s when Germany was saddled with a massive debt from the costs of the First World War and subsequent reparations that it agreed to pay. By November 1923, the US dollar would have bought you 4.2 trillion German marks, and the currency was so worthless that banknotes were being used as wallpaper or kindling, and workers took suitcases and wheelbarrows to work to collect their wages. The following July, a total of 1.2 sextillion – 1,200,000,000,000,000,000,000 – marks were in circulation.

Just after the Second World War, Hungary suffered a similar financial crisis when its inflation rate hit 41.9 quintillion per cent – the highest on record. This prompted the issue of banknotes of ridiculously large denomination, culminating

FIGURE 2.1: In the Weimar Republic, in the 1920s, stamps were issued and restamped in response to the rapid inflation of the German mark.

in the 1946 Százmillió-B pengő. 'Százmillió-B' stands for 100 million million million, so with one of these in your pocket you'd have been the proud owner of a hundred quintillion (10^{18}) pengős, enough to buy – well, probably not very much! To put it in context, assuming a single pengő note measured 5 centimetres by 10 centimetres, 100 quintillion of them would have been enough to completely cover the Earth's surface 2,000 layers deep. Things got so bad in Hungary that it was decided to introduce an entirely new currency. On 1 August 1946, the forint replaced the pengő, or, to be more accurate, 1 forint replaced 400 octillion, or 4×10^{29}, pengős! At a stroke, the total amount of pengő notes in circulation was reduced in value to less than one thousandth of a single unit of the new currency.

Still on the subject of finance, the richest person on Earth is Jeff Bezos, founder of Amazon, with an estimated wealth of $190 billion. But that's peanuts compared to the largest ever lawsuit, for two undecillion (two trillion trillion trillion) dollars, filed on 11 April 2014, by Manhattan resident Anton Purisima after claims that he was bitten by a 'rabies-infected' dog on a New York City bus. In a rambling 22-page handwritten complaint, accompanied by a photo of an unreasonably outsized bandage around his middle finger, Purisima sued NYC Transit, LaGuardia Airport, Au Bon Pain (where he insisted he was routinely overcharged for coffee), Hoboken University Medical Center, and hundreds of others for vastly more money than there is on the planet. Fortunately for the defendants – and the world's monetary supply – the judge dismissed the lawsuit, stating: 'The Court, after reviewing Plaintiff's complaint, finds that it lacks any arguable basis in law or in fact.'

As you can imagine, some of the biggest numbers of relevance to physical reality come about when we try to count how many of the smallest things ever conceived by science would fit into the largest thing of which we know – the universe itself. But it's not immediately obvious what are the values of these two extremes in size. The entire universe is thought to have sprung into existence in an event known as the Big Bang some 13.8 billion years ago. Ever since then the cosmic contents have been flying apart like debris from a colossal explosion, scattered further and further asunder on the spreading surface of four-dimensional space-time.

The overall nature of the universe, including its shape and whether it's finite or infinite in size, remains uncertain.

We can get information only from the part we can see: the portion of the universe from which light has had a chance to reach us since the Big Bang. This is the so-called *observable* universe, which is believed to have a diameter of about 93 billion light-years or 8.8×10^{26} metres. If we assume that it's spherical in shape then its volume comes out to be roughly 4×10^{80} cubic metres.

At the other end of the scale is the world of atoms, subatomic particles and quantum mechanics. A hydrogen atom, for instance, measures just one ten billionth of a metre across. That's one divided by ten billion – one over 10^{10} – which we can write as 10^{-10} metres, the minus sign indicating the 'one over' or reciprocal. A hydrogen nucleus, a single proton, is far smaller, with a diameter of just 2×10^{-14} metres and a volume of 4×10^{-42} cubic metres. In terms of volume, then, the observable universe is 10^{122} times bigger than a proton.

But a proton is a long way from being the smallest thing in nature. Inside the proton are even tinier particles called quarks. And much smaller than any fundamental piece of matter is something known as the Planck length. Physicists hypothesise that, viewed at sufficiently high magnification, the graininess of spacetime itself would become evident. Normally, we think of spacetime as being a smooth and continuous backdrop against which the drama of matter and energy plays out. But at short enough lengths and times, this continuity is thought to break down and the quantum nature of space and time become apparent. Named after German theoretical physicist Max Planck, who first proposed it in 1899, the Planck length is about 1.6×10^{-35} metres. A sphere with this diameter would have a volume of about 2×10^{-105}

cubic metres. Dividing the current volume of the observable universe (4×10^{80} cubic metres) by this fantastically tiny number tells us how many times the smallest physically meaningful volume will fit into the largest actual volume of which we know. The answer: about 2×10^{185}. Written out in full (with apologies to the proofreader!) this is

200,000,000,000,000,000,000,000,000,000,000,
000,000,000,000,000,000,000,000,000,000,000,
000,000,000,000,000,000,000,000,000,000,000,
000,000,000,000,000,000,000,000,000,000,000,
000,000,000,000,000,000,000,000,000,000,000,
000,000

or 200 sexagintillion.

We don't know how much bigger the entire universe is than the observable universe. One possibility is that the universe is, and always has been, infinite in size. Even if it's finite, the part we can see is likely to be only a small fraction of the whole. Much depends on what happened during an extraordinarily rapid, early phase of cosmic evolution, known as inflation, which began about 10^{-37} seconds after the Big Bang. Lower estimates suggest that the whole universe has at least 250 times the volume of what's currently observable. Higher estimates put it in excess of 10^{70} times as big. In any event, the ratio of the volume of the universe in its entirety to the volume spanned by a single Planck length must be far greater than the already-gigantic number we wrote out in full above.

Colossal numbers also arise when we start to think about the limits of computation in the real world. Here we're talking not about what, in theory, it's possible to compute given endless amounts of time and storage space (more on this later), but instead the practical limits imposed by the laws of physics. One of these is known as Bremermann's limit after the German-American mathematician and bio-physicist Hans-Joachim Bremermann. It's based on two fundamental principles. The first is the equivalence of mass and energy as specified by Einstein's famous relation $E = mc^2$, where c is the speed of light. The second is Heisenberg's uncertainty principle, which expresses the precision with which certain pairs of quantities, such as energy and time, and mass and momentum, can be known. Bremermann's limit is the maximum rate at which data can be processed by any isolated material system. It's equal to c^2/h, where h is Planck's constant, and comes out to be about 1.36×10^{50} bits per second per kilogram. We're not used to seeing processing capacity expressed this way. Normally, we read about how many bits per second a chip or a computer as a whole can process. That's because Bremermann's limit far exceeds the processing rate of anything humans have ever made or are likely to make in the foreseeable future. It is, however, of interest in the design of cryptographic systems that are secure against brute-force searches that try to crack secret codes and passwords by running through all possible combinations.

Imagine a computer as massive as the Earth that can work at Bremermann's limit. It would be capable of roughly 10^{75} computations every second. At this rate, it could break a typical 128-bit cryptographic key in less than 10^{-36} seconds and a 256-bit key in about two minutes. However, cracking a 512-bit key, even for a planet-sized behemoth working at the computational speed limit, would take an impossibly long 10^{72} years.

Another (and related) extreme limit on what's computationally possible in nature is the Bekenstein bound, named after Mexican-born Israeli-American physicist Jacob Bekenstein who came up with the idea in the early 1980s. It refers to the maximum amount of information that can be contained within a given volume of space. In defining his bound, Bekenstein had in mind the most extreme objects in the universe – black holes. Specifically, he was looking into the problem of what happens when something falls into a black hole.

It used to be thought that, once having entered a black hole, nothing could ever escape. However, Stephen Hawking proposed a way that information might be extracted, leading to the possibility of using a black hole as the most extreme data storage or computational device, with a storage density equal to the Bekenstein bound. MIT engineer and physicist Seth Lloyd calculated that the 'ultimate laptop', formed by compressing a kilogram of matter into a black hole 3×10^{-27} metres across would be able to perform 5×10^{50} operations per second. On the downside, being a submicroscopic black hole, it would evaporate in a flash of gamma rays, due to Hawking radiation, in about a ten million trillionth of a second – making it less than ideal for everyday use!

At the other end of the scale, Lloyd figured, if all the matter in the observable universe were turned into a black hole computer, it would be capable of 10^{90} operations per second and have a lifetime of 2.8×10^{139} seconds before Hawking radiation caused it to evaporate. In that time it would be able to perform 2.8×10^{229} operations – though given that the computer would be coextensive with all of physical reality, it's hard to know what the purpose and nature of those operations would be!

In the worlds of science fiction anything is possible, even being pushed out of an airlock of one spaceship and being rescued moments later by another spaceship that just happens to be passing by. In *The Hitchhiker's Guide to the Galaxy*, Douglas Adams gives the odds of his characters Arthur Dent and Ford Prefect enjoying such a monumental slice of luck, after being ejected from a Vogon spacecraft: 2^{260199} to one against. As it happens, they *are* rescued – by *The Heart of Gold*, powered by an 'infinite improbability drive'.

It might not seem possible that there could be numbers even bigger than this that describe aspects of the *real* world in which we live. But there are some that are ridiculously larger than all of those we've looked at so far – and, as in Adams' book, they involve probability. These numbers refer to what are known as Poincaré recurrence times and they're spectacularly stupendous in size. A Poincaré recurrence time, named after French mathematician and physicist Henri Poincaré, is the time taken for a system, such as a collection of moving particles, to return to exactly the same state in which it started. Naturally, the bigger and more complex the system, the more possible states it can have and the longer it will take to come back, purely by chance, to its initial condition. Canadian physicist Don Page has calculated Poincaré recurrence times for the entire universe under a variety of different starting conditions. Depending on the model used to describe the earliest moments of the universe, Page came up with a range of recurrence times extending all the way up to this many years:

$$10^{10^{10^{10^{10^{1.1}}}}}$$

Nothing we've talked about so far comes anywhere near this in size. In fact, it's the largest number ever to be calculated and published in an academic journal by a scientist (though not by a mathematician!). It's also the first time, in this book, that we've seen something represented in this way, in the form of a stack of exponents: 10 to the 10 to the 10 to the 10 to the 10 to the 1.1. For convenience, we can also write it, using carets or circumflex accents, as 10^10^10^10^10^1.1.

We'll be seeing a lot more of this type of number – a repeated exponential – later on. For now, it's worth pausing a moment to understand what it means. Ignore the 1.1 and think about 10^10^10^10^10. Evaluating this from right to left – in other words, going down the powers – 10^10 is 10,000,000,000; 10^10^10, is 10^10,000,000,000 or one followed by 10 billion zeros; 10^10^10^10 is 10 *to the power of* one followed by 10 billion zeros; and, finally, 10^10^10^10^10 is 10 raised to the power 10 to the 1 followed by 10 billion zeros.

This might seem a seriously impressive number but it's just a baby step along the road to the biggest ones we can name, define, or contemplate. From now on, we must start to leave behind the physical world and enter the domain of mathematics, where the mind is unshackled from limitations of matter, energy, space, and time.

CHAPTER 3

Maths Unbound

SCARILY LARGE NUMBERS can quickly grow from humble beginnings, as the old wheat-and-chessboard problem demonstrates. The earliest recorded version of this appears in the writings of Islamic scholar Ibn Khallikan in 1256. According to the tale, King Shirham of India wanted to thank Grand Vizier Sissa ben Dahir for his invention of the game of chess and asked Sissa what he'd like as a reward. Sissa's request seemed pitifully modest: one grain of wheat for the first square of the chessboard, two for the second, four for the third, and so on, the amount doubling each time until all the squares had been covered. The King, thinking he'd got off lightly, immediately granted the wish but then soon regretted it. The total amount of grains needed turns out to be $1 + 2 + 2^2 + 2^3 + \dots + 2^{63} = 18{,}446{,}744{,}073{,}709{,}551{,}615$ – a number far greater than all the wheat grains in the world.

King Shirham learned a couple of lessons from his encounter with the crafty Grand Vizier. First, he found out about the surprising power of repeated doubling or, more generally, that of exponentiation – in which a number is multiplied by itself over and over again. And second, it was brought home to him that maths isn't confined by practical considerations. Physical limits don't restrict the world of the mathematician. On the contrary, this world is big enough to contain anything and any number we can imagine.

Among the most interesting and important numbers in maths are prime numbers. Their distinguishing feature is that they're divisible only by themselves and one. The first ten of them are: 2, 3, 5, 7, 11, 13, 17, 19, 23, and 29. Gradually, as they get larger, the gaps between them tend to widen. For example, after 1000, the first ten are 1009, 1013, 1019, 1021, 1031, 1033, 1039, 1049, 1051, and 1061. It's easy to prove that there are infinitely many prime numbers. But what particularly interests mathematicians is how they're distributed – the details of the way in which they thin out at larger and larger values. One of the greatest unsolved problems in maths, the Riemann hypothesis, is intimately concerned with this very issue. There are also practical reasons for seeking out bigger and bigger primes. Most importantly, very big primes are crucial to the most widely used systems for encrypting data, such as those that underpin all online banking and shopping.

The easiest-to-find primes are those that happen to be Mersenne numbers, named after French friar and polymath Marin Mersenne, who studied them in the first half of the seventeenth century. All Mersenne numbers can be expressed in the form $2^n - 1$, where n is a positive integer: in other words, they are one less than the corresponding power of 2. The first few are 1, 3, 7, 15, 31, 63, and 127. For small values of n, whenever n is prime, the corresponding Mersenne number is also prime. For example, when $n = 7$,

$2^n - 1 = 127$, which is prime because it has no factors other than 1 and 127. But this pattern breaks down for $n = 11$, because $2^{11} - 1 = 2047 = 23 \times 89$. Mathematicians now know that although in the case of every Mersenne prime n must also be prime, other conditions have to be satisfied as well. Fortunately, these extra conditions are easy enough to be coded into a program so that computers can crunch their way through relatively simple, fast algorithms in the quest for the next biggest Mersenne prime.

The process of finding them can be used as a way of testing the speed and capability of new, faster computers and algorithms – the step-by-step methods used to solve problems. It also comes in handy as a marketing ploy for these new machines, because the appearance of a new, largest known prime tends to make front-page news. One of us (David) has some personal experience of this.

In the late 1970s, I worked for the Minneapolis-based supercomputer manufacturer Cray Research in the applications development group. One of our tasks was to show how much quicker the Cray-1 – then the world's fastest computer – was than any of its rivals. In this we were helped by the efforts of a young software engineer, David Slowinski. I had the opportunity to meet David on a number of occasions and he explained to me how his Mersenne prime-hunting algorithm worked. He and Harry Nelson, a mathematician at the Lawrence Livermore National Laboratory, where one of the earliest Cray-1s was undergoing acceptance tests, had optimised the algorithm to run on the Cray's unusual 'vector' architecture.

In April 1979, Slowinski and Nelson's efforts were rewarded with the discovery of the 27th Mersenne prime,

$2^{44,497} - 1$, the largest prime number known at the time. This also gained them a place in *The Guinness Book of Records*. Slowinski went on to discover six more record-breaking primes between 1982 and 1996, culminating in $2^{1,257,787}$, or M(1257787), the 34th Mersenne prime, which he found, in collaboration with computer scientist Paul Gage, using a Cray T90 supercomputer.

Of the ten largest prime numbers found to date, nine are Mersenne primes. The current champion, at the time of writing, discovered in January 2018, is $2^{82,589,933} - 1$. Written out in full it has close to 25 million digits – enough to fill several volumes of the last print edition of *Encyclopaedia Britannica*.

Mersenne primes can be used to find another kind of interesting number known as perfect numbers. A perfect number is a whole number that is equal to the sum of all its factors except itself. For example, 6 is a perfect number because its factors, 1, 2, and 3, add to give 6. The next smallest is 28 (the sum of $1 + 2 + 4 + 7 + 14$). In about 300 BCE, Euclid proved that whenever $2^n - 1$ is prime (where n itself is prime) – in other words, for every Mersenne prime – $2^{n-1}(2^n - 1)$ is an even perfect number. For example, the first four perfect numbers are generated like this:

For $n = 2$: $2^1(2^2 - 1) = 2 \times 3 = 6$
For $n = 3$: $2^2(2^3 - 1) = 4 \times 7 = 28$
For $n = 5$: $2^4(2^5 - 1) = 16 \times 31 = 496$
For $n = 7$: $2^6(2^7 - 1) = 64 \times 127 = 8128$

All even perfect numbers end in 6 or 8 and have a corresponding Mersenne prime. There are no known odd perfect numbers. So, up to the present, there's a one-to-one correspondence between known Mersenne primes and known perfect numbers. The largest known perfect number is that corresponding with the largest known Mersenne prime. It has the value $2^{82,589,932}(2^{82,589,933} - 1)$. Written out in full it involves just under 50 million digits, so, to avoid losing readers, we'll just show how it starts and finishes: 110847779864...007191207936.

Strangely enough, although we know that there are infinitely many primes, we don't know if there are infinitely many Mersenne primes. Likewise, we don't know if there are infinitely many perfect numbers or if there are any odd perfect numbers.

In any event, it's easy to come up with a number that's bigger than the biggest known prime or perfect number: just add 1! But large numbers that have some feature of interest other than being simply big hold a special fascination. Sometimes the large number in question marks the earliest point at which some mathematical condition is satisfied. This is the case with Skewes' number, named after South African mathematician Stanley Skewes. Like the Riemann hypothesis, to which it's related, Skewes' number has to do with the distribution of primes.

Born in the Transvaal to an American-born mother and an English father in 1899, Stanley Skewes took a degree in civil engineering at the University of Cape Town before heading to Cambridge to study mathematics. At King's College, where among his contemporaries and rowing partners was Alan Turing, he studied for his BA, MA, and finally PhD. His doctoral supervisor, John Edensor Littlewood, was an eminent number theorist and close collaborator of G. H. Hardy (who brought the Indian genius Ramanujan to England). It was Littlewood who encouraged Skewes to pursue research into prime numbers.

Mathematicians had searched long and hard for a formula that would generate prime numbers, without success. However, they had found that the occurrence of primes isn't haphazard; rather, there are rules governing their distribution. Their density decreases when moving to larger and larger numbers. A result known as the prime number theorem was first proposed by German mathematician Carl Gauss in about 1792 when he was just a teenager. However, it wasn't until 1896 that it was proved, independently by French mathematicians Jacques Hadamard and Charles de la Vallée Poussin. According to this theorem, the density of primes around any number x is approximately $1/\log_e x$, where $\log_e x$ is the natural logarithm of x. So, around 100 we'd expect about 1 in 5 numbers to be prime, whereas around 1,000 this falls to about 1 in 7.

The prime number theorem makes it possible to estimate how many primes there are less than any given number n. In fact, for large values of n, the value predicted by the theorem is remarkably close to the actual number of primes. For example, there are exactly 50,847,534 primes less than one billion, whereas the theorem predicts a value of 50,849,235 – just 1,701, or 0.0033 per cent, more. The estimated value from the prime number theorem is consistently a little bit higher than the actual number of primes, even when n is enormous. In the absence of evidence to the contrary, mathematicians assumed this would always be the case, without exception. But then Littlewood came up with a proof that, at some point, the overestimates would stop and underestimates would take over for a while, before there was a switch back to overestimates, and so on, back and forth, forever. Littlewood didn't know when the first crossover would happen. However, his student, Skewes, managed to shed some light on this question. Skewes showed that the prime number theorem must flip to giving underestimates before n reaches the value 10^10^10^34. This spectacularly big number – Skewes' number – assumes that the Riemann hypothesis is true. The Riemann hypothesis effectively says that prime numbers are just about as organised and predictable as it's possible for them to be given their nature. Skewes calculated that if the hypothesis were false then the prime number theorem would first switch to underestimating the number of primes when n was even larger – about 10^10^10^964, which became known as Skewes' second number.

G. H. Hardy once described Skewes' number as 'the largest number which has ever served any definite purpose in mathematics'. Although it's long since lost that distinction, its immense size does underscore an important point: that just because something has held true as far as anyone has managed to test doesn't guarantee it'll be true in every conceivable case. This is one of the reasons mathematicians are never happy until they've found a rigorous proof that will stand for all time.

Since Skewes published his results, the lower and upper bounds (depending on whether the Riemann hypothesis is assumed to be true or not) have been drastically reduced. For the past half-century or more, computers have been used in these calculations. In 1966, American mathematician Sherman Lehman showed that somewhere between the values for n of $1.53 \times 10^{1,165}$ and $1.65 \times 10^{1,165}$ there are more than 10^{500} consecutive integers for which the prime number theorem underestimates the actual number of primes up to that point. No one's yet identified a specific number where the crossover first happens. The most recent work published on the subject, by Stefanie Zegowitz of the University of Manchester in 2010, found that the first time the prime number theorem gives an underestimate is for some value of n that's less than 1.3972×10^{316}. Although still big by everyday standards, this value is minuscule compared with the original Skewes' number.

Other gigantic numbers in maths emerge from problems to do with combinations and probability. Take an ordinary pack of playing cards. How many different ways can the deck be arranged? The first card can be any of 52, the second any of the remaining 51, and so on, so that the total number of possible arrangements is $52 \times 51 \times 50 \times \ldots \times 3 \times 2 \times 1$. This is known as 52 factorial, or 52!. Written out in full it turns out to be:

80,658,175,170,943,878,571,660,636,856,403,766,
975,289,505,440,883,277,824,000,000,000,000

or, in scientific notation, approximately 8.0658×10^{67}. If a deck had been randomly shuffled every second since the start of the universe, there would have been time for only about 4.32×10^{17} shuffles – an insignificant proportion of all the different ones possible. It's fair to say that any story about a random shuffle having produced a perfect ordering of a deck isn't true – the odds are 80 million trillion trillion trillion trillion trillion to one against! Such claims generally mean that someone is fibbing or the shuffle wasn't truly random.

Chess offers even more possible combinations. American mathematician and electrical engineer Claude Shannon, often referred to as the father of information theory, was fascinated by the link between chess and the emerging new field of computer science. On 9 March 1949, he presented a paper called 'Programming a Computer to Play Chess' at the National Institute for Radio Engineers Convention in New York. In this he argued that the number of possible positions that any (legal) number of chess pieces could be in at a given point in a chess match was between 10^{43} and 10^{50}. He also estimated that the number of possible games two players could play without repeating themselves – the so-called Shannon number – was around 10^{120}. Of course, this includes all kinds of games that would never happen in practice unless one or both players hadn't a clue what they were doing. The number of *realistic* games between players who have at least a rudimentary knowledge of the game and manage to avoid ridiculous move combinations is much lower but still immense. The Shannon number or, more formally, the *game-tree complexity*, of chess is so huge, Shannon realised, that for computers to be successful in competing with humans their programs couldn't be based simply on analysing every possible position from the outset.

Other familiar games have their own game-tree complexities. Tic-tac-toe, also known as noughts and crosses, is obviously much simpler than chess. Nevertheless, it can give rise to a surprisingly large number of game situations. The first player can put a 'O' or 'X' in any one of nine places. The second player can choose from the eight remaining places, and so on. Continuing this line of reasoning gives a value of 9! = 362,880 for the upper bound of the game-tree size. But this number includes many illegal situations, since the rules say that the game ends as soon as one player achieves a line of three 'O's or 'X's, so that not all nine squares have to be filled in. The number of possible legal games turns out to be 255,168, or a mere 26,830 if reflections and rotations of positions are treated as being identical.

The popular game of Connect Four is much more complicated than tic-tac-toe. It's played using a 6 up by 7 across vertical board and two sets of coloured counters. The winner is the first to make a line of four counters of their colour. Altogether there are 4,531,985,219,092 different board positions and about 1.1×10^{20} possible game situations. Backgammon is much more complex than Connect Four. In fact, the number of possible games of backgammon is about 10^{144} – far greater than the number of possible chess games, even though there are many more possible board positions in chess than backgammon. A vastly more complicated game than any we've yet mentioned is Go. Invented in China some two and a half thousand years ago, it's most commonly played on a square grid made up of 19 × 19 lines. Two players, black and white, take it in turn to place stones of their colour at the intersections of the lines. Although the rules are simple, the strategy needed to play well is fiendishly hard. Go is estimated to allow about 10^{172} board positions and about 10^{360} different games.

If the universe is considered to be made up of many tiny cells, or subatomic-sized locations, each of which can be in a variety of different states then we can think of it as being like a colossal board game. This brings us back to the concept of Poincaré recurrence times, which we talked

FIGURE 3.1: A game of Go in progress. With a standard 19 × 19 board, it's estimated that there are 10^{360} possible different games.

about in the last chapter. In particular, if the cosmic game is played on a board with cells that are a Planck unit wide, and with the smallest particles in nature, then the time, in years, needed for the board position to return to a particular state is the vast number we mentioned that was calculated by Don Page – 10^10^10^10^10^1.1.

This problem of figuring out the probability of a complicated system reaching a specific state has cropped up in various guises over the centuries. A classic thought experiment, dating back more than a century, involves monkeys and typewriters. It was discussed as early as 1913 by French mathematician Émile Borel, and perhaps by others even before that. A common statement of it is: how long would it take a monkey, hitting the keys of a typewriter at random, to type out *Hamlet*, word for word, exactly as Shakespeare had written it? Obviously, most of what the monkey would type would be utter gibberish. Even coming up with a valid three-letter English word would be a rarity.

The first line of the play, spoken by Bernardo, is: 'Who's there?' The chances of the monkey hitting the 'W' are 1 in 44 because there are 44 keys on a standard typewriter keyboard. The chances of hitting the 'H' immediately after the 'W' are also one in 44, so that the chances of hitting the sequence 'WH' are one in 44 times 44, or one in 1936. The odds of our hypothetical simian getting the first word of *Hamlet* correct on any given attempt are a pitiful one in 44 times 44 times 44, or one in 85,184. Next it has to hit the apostrophe, then the 'S', then a space. You see the problem: you'd have a better chance of winning EuroMillions – and we haven't even started on the second word! What's more, if the monkey makes a single mistake, all its efforts are undone and any correct sequence it's managed to produce up to that point has to be thrown out and a fresh start made.

Hamlet, the longest of Shakespeare's plays, contains 30,557 words and approximately 130,000 letters, punctuation marks, and spaces between words. The odds of typing it correctly in one go, using random keystrokes, are about one in $4.4 \times 10^{360,783}$, assuming that capital letters are separate keys. On an actual typewriter, capital letters and some punctuation marks require the simultaneous pressing of two keys (the shift key plus one other), which vastly diminishes the probability of success. The fact is that even if the universe were crammed full of subatomic-sized monkeys that had been tapping away non-stop on subatomic-sized typewriters since the Big Bang, there'd be only a one in a trillion chance of them typing even the first 79 characters of *Hamlet* correctly.

Of course, real monkeys aren't inclined to spend their days working diligently, without break, as typists, even of the hit-or-miss variety. This fact was borne out when, in 2003, as a piece of performance art, a computer keyboard, connected by a radio link to a website, was left in the enclosure of six Sulawesi crested macaques at Paignton Zoo in Devon. The animals could watch the results of their efforts on a screen. While the novelty of their new plaything lasted, the monkeys managed to produce six pages of writing, consisting mostly of the letter 'S' before, in a fit of rage, the alpha male attacked the keyboard with a stone and the others ended up urinating on it.

In 2011, American programmer Jesse Anderson enlisted the help of an army of virtual, and more compliant, monkeys to recreate not just *Hamlet* but all of Shakespeare's works. Using Amazon's EC2 cloud computing system, he ran millions of small computer programs simultaneously, each churning out random sequences of nine characters. However, it was a much watered-down version of the original monkeys-and-typewriters challenge. Anderson required only that a nine-letter sequence appear anywhere in the Bard's writings for it to be checked off as completed. It doesn't compare with the feat of having to get all the letters and other characters correct for *Hamlet* in one go. On the upside, as Anderson pointed out: 'This is the largest work ever randomly reproduced. It is one small step for a monkey, one giant leap for virtual primates everywhere.'

A similar concept to that of the monkeys and typewriters has been explored by writers such as Jorge Luis Borges. In his

1941 short story *The Library of Babel*, Borges imagines an immense library containing a vast number of books, identical in size and format: 'each book contains four hundred and ten pages; each page, forty lines; each line approximately eighty black letters.' Only 22 alphabetic characters plus a comma, full stop, and space are used throughout, but every possible combination of these characters that follow the common format occurs in some book in the library. Most books appear to be just a meaningless jumble of characters; others are quite orderly but still devoid of any apparent meaning. For example, one book contains just the letter 'A' repeated over and over. Another is exactly the same except that the second letter is replaced by a 'B'. Others have words, sentences, and whole paragraphs that are grammatically correct in some language but are nevertheless illogical. Some are true histories. Some purport to be true histories but are, in fact, fictional. Some contain descriptions of devices yet to be invented or discoveries yet to be made. Somewhere in the library is a book that contains every combination of the basic 25 symbols that can be imagined or written down in the given format. Yet, all of it's useless because without knowing in advance what's true or false, fact or fiction, meaningful or meaningless, such exhaustive combinations of symbols have no value.

How big would Borges' library be? Jonathan Basile, of Emory University, has simulated an English-language version of it on the website *LibraryofBabel.info*. An algorithm he wrote generates a 'book' by iterating every permutation of 29 characters: the 26 English letters, plus space, comma, and full stop. Each book is marked by a coordinate, corresponding to its place in the hexagonal library, so that every book can be found in the same location every time. The website is said to contain 'all possible pages of 3,200 characters, about $10^{4,677}$ books'. The site also has a search tool with which users can find the whereabouts in the library of any known page of text. An individual page of *Hamlet* can be found in this way, though the probability of finding any other page from the same work in the same volume is next to zero. If a user clicked through the books at a rate of one per second, it would take about $10^{4,668}$ years to go through the library.

Essentially useless though it may be, there's something compelling about Basile's creation. It contains every page that has ever been written or ever will be written (within the limits of its format) – every news story, past, present, and future, every play and novel, every work of fact and fiction (including this book), and every scientific discovery that will ever be made. Somewhere within it, along with much that is meaningless or false, lie accurate descriptions of every planet across the vast reaches of space and time, and the true details of how life and the universe itself originated.

We've already come across some impressively large numbers and it's time to start thinking about how we might represent even larger ones in a convenient, manageable way. The only mathematical operations we normally use for dealing with big numbers are addition, multiplication, and exponentiation. Addition is just repeatedly adding 1 to some starting number (in other words, counting up one at a time). If the starting number is 8 and we want to add 4 to it, we can write this as 8 plus four 1s: $8 + 4 = 8 + 1 + 1 + 1 + 1 = 12$. Multiplication is repeated addition, so that, for instance,

$7 \times 4 = 7 + 7 + 7 + 7 = 28$, and exponentiation is repeated multiplication: for example, $3^6 = 3 \times 3 \times 3 \times 3 \times 3 \times 3 = 729$. For most practical purposes, we don't need any operation that's more powerful or compact in its notation than raising one number to the power of another. But we're about to venture into realms that go far beyond the practical so we'll need something that transcends our familiar ways of representing numbers.

Let's start by writing down the operations of addition, multiplication, and exponentiation in a general form. In fact, let's start right at the beginning with something we don't often hear about – the *successor* operation. This gets us off the ground, mathematically speaking, starting from zero and adding on one and then one more, and so on. The simplest thing we can do is add one on to a number a to give the successive number:

$a + 1$

The successor operation (at least in the form it's defined here) is the first thing we learn in mathematics: how to count up, one at a time. The next step up is general addition, in which we repeatedly apply the successor operation, in other words, begin with a number a and add on b ones. We can write this as:

$a + b = a + (1 + 1 + \ldots + 1)$, where there are b copies of 1 in the brackets.

Next up is multiplication. If we're multiplying a times b, we can set this out as:

$a \times b = a + a + \ldots + a$, where b copies of a are added together.

Finally, we can write a^b (a to the power of b) as:

$a^b = a \times a \times \ldots \times a$, where b copies of a are multiplied together.

Notice that each level of operation can be expressed in terms of the one immediately below (for example, multiplication in terms of repeated addition) and, in fact, is just a compact way of thinking about and representing this lower operation.

For most purposes we don't need to go beyond the a^b (exponential) stage. The kind of big numbers we meet in everyday life, read about in the news, or, for that matter, with which most scientists and mathematicians ever deal, can be represented comfortably in exponential form. The two richest people in the world, at the time of writing, are Jeff Bezos of Amazon, with about \$190 billion to his name, and Bill Gates of Microsoft, with about \$120 billion. That's 'only' one or two times $\$10^{11}$ each! All the money in the world can be written very compactly as about 37×10^{12} US dollars. Even physicists and astronomers, studying everything from subatomic particles to the cosmos as a whole, can manage perfectly well with numbers in exponential form.

We've come across a few numbers in this book, though, which are too big to be written simply as one number to the power of another. The googolplex, for instance, is 10^{googol} but writing it this way is cheating since a googol equals 10^{100}. We should really write it as:

$$googolplex = 10^{10^{100}}$$

using a repeated exponential. In the same way, if we think back to Skewes' number and Page's cosmic recurrence time, these need a stack of powers in order that we can represent them without filling this book, or the entire universe, with zeros.

Carets or circumflex accents, ^, are commonly used to indicate powers of powers and are a handy device, especially if you want to avoid typesetting problems. The trouble is, once you start dealing with numbers that are far bigger than the googolplex or Skewes' number, or even Page's cosmic recurrence time, it becomes unwieldy to show them like this. Just as it's much easier to read and write one hundred trillion trillion as 10^{14} than as 100,000,000,000,000, so we'd like a more concise way of representing a gigantic number like, say 10^10^10^10^10^10^10^10^10^53.

English mathematician Reuben Goodstein, in his 1947 paper 'Transfinite Ordinals in Recursive Number Theory', coined the name 'tetration' for the operation that was the next step up from exponentiation. The word combines the Greek *tetra* for 'four' with 'iteration', and conveys the fact that tetration is the fourth level in the hierarchy of operations after addition, multiplication, and exponentiation.

Just as we can summarise exponentiation as

$a^b = a \times a \times \ldots \times a$, where b copies of a are multiplied together.

So we can summarise tetration as

$$^b a = \underbrace{a^{a^{\cdot^{\cdot^{a}}}}}_{b}$$

where b copies of a are combined by exponentiation, working from right to left.

For instance, $^5 10$ means 10^10^10^10^10 or

$$10^{10^{10^{10^{10}}}}$$

Others, such as Nick Bromer of Bryn Mawr College in a paper published in 1987, have used the term 'superexponentiation' instead of tetration. But the two words mean exactly the same thing: both refer to the next operation in the sequence: addition, multiplication, and exponentiation. Most of us never come across superexponentiation or tetration. Neither, for that matter, do most mathematicians unless they happen to specialise in a field that involves having to deal with fantastically large numbers. But since this is a book specifically about big numbers, we'll need to get used to these strange terms. Coming up with meaningful names for big numbers, and ways of defining them, is part of the challenge of the subject we're talking about here.

We also need to be careful about getting our terminology accurate. For instance, Skewes' number

$$10^{10^{10^{34}}}$$

isn't a tetration because not all the exponents are the same. Instead, like the googolplex and Page's cosmic recurrence time, it takes the form

$$a^{a^{\cdot^{\cdot^{a^x}}}}$$

which is called an iterated exponential. It's also possible to have a number that looks like

$$a_1^{a_2^{\cdot^{\cdot^{a_n}}}}$$

where all the exponents are different, in which case it would be called a nested exponential.

As we'll see later, there are other ways of notating tetrations that allow an easy extension to even more powerful operations. After all, why stop at tetration? If tetration is repeated exponentiation then repeated tetration brings us to the next level of operator, which Goodstein called pentation. Next would be hexation, and so on. Mathematicians refer to this sequence of arithmetic operations – succession, addition, multiplication, exponentiation, tetration, pentation, and so on – as the hyperoperation sequence – and it goes on forever. Starting from succession, which is called hyper0, tetration is also known as hyper4, pentation as hyper5, hexation as hyper6, and so on.

Although we mentioned Reuben Goodstein as having invented the name 'tetration', he certainly wasn't the first to deal with the concept or the nomenclature of hyperoperators. The notation ^na for a raised to the a n times was introduced by German mathematician Hans Maurer in 1901, although it didn't come to widespread attention until 1995, when Rudy Rucker wrote about it in his book *Infinity and the Mind*.

As for the origin of ideas about operators more powerful than exponentiation, we have to go further back in time. The story really begins with Johann Lambert, an eighteenth-century Swiss polymath who made an extraordinary number of important contributions to maths, physics, astronomy, logic, and philosophy. He invented the first practical hygrometer (for measuring humidity), came up with the nebular hypothesis for the origin of the Solar System (an ancestor of the model accepted today), suggested (correctly) that the Sun is part of a group of stars that moves collectively through our galaxy, and did pioneering work in photometry. In maths, he gave the first rigorous proof that pi is irrational, investigated the general properties of map projections, introduced hyperbolic functions into trigonometry, and speculated about non-Euclidean geometry.

Lambert also defined something that's come to be known as the Lambert W-function, otherwise called the omega function or product-logarithm. He used it to solve problems that couldn't be solved, at the time, by any conventional means available. (An example would be solving the equation $3x = 3^x$.) There's no need to go into the intricacies of what the W-function involves other than to say that it laid the basis for figuring out what happens when a number is raised to the same power over and over again indefinitely. Lambert wrote about this function in his first book, oddly enough on the subject of light passing through various media, published in 1758. However, it was his compatriot Leonhard Euler who used the W-function in a way that links it to tetration via what's become known as the power tower function.

Euler was twenty years older than Lambert and among the most prolific mathematicians in history. In one of his papers he referred to 'the ingenious engineer Lambert', which sounds a bit like a back-handed compliment! Their mathematical interests overlapped, especially in areas of number theory. Euler proved that e was irrational before

Lambert proved the same about π. Lambert's work on the W-function led Euler to consider the properties of what became known as the power tower function:

$$y = f(x) = x^{x^{x^{x^{\cdot^{\cdot^{\cdot}}}}}}$$

in which the xs are stacked infinitely high. A quick glance might suggest that for *all* values of x greater than one this function will shoot off to infinity. We're so used to using the term 'exponential growth' to mean 'explosive growth' that it seems that a repeated exponential like Euler's should just blow up without limit whenever $x > 1$. But, remarkably, this isn't the case: there are many values of x, greater than one, for which Euler's power tower function converges to a finite value. For instance if $x = \sqrt{2}$, then:

$$y = \sqrt{2}^{\sqrt{2}^{\sqrt{2}^{\sqrt{2}^{\sqrt{2}^{\cdot^{\cdot^{\cdot}}}}}}}$$

and the value of this, it's easy to prove, isn't infinity or some gigantic finite number, but just 2. In investigating his power tower function, Euler became a pioneer of hyperoperations that lie beyond simple exponentiation. He was the first to look in depth at what we now call tetration and, although he recognised that raising a power to a power to a power and so on could quickly lead to colossal numbers, he also turned up some surprises. The path to the biggest number in the world isn't just relentlessly upward but has fascinating twists and turns.

CHAPTER 4

Up, Up and Away

TECHNOLOGICAL PROGRESS comes in fits and starts. For a while, not much happens beyond a steady development of what's already known or been invented. Then, all of a sudden, comes a breakthrough that whisks us up to an entirely new level. This happened, for instance, with the invention of the practical steam engine at the start of the Industrial Revolution. It happened again with computers, when mechanical devices were replaced by electronic ones, and then again, though less visibly to the public at large, when vacuum tubes (or valves) were superseded by transistors. It's the same in mathematics: a moment of inspiration can lift us into a realm of thought that we never suspected existed.

When it comes to big numbers – *really* big numbers – one of the trickiest problems is how to represent them. We quickly get tired of writing long rows of zeros and welcome being able to show a big number in terms of powers of ten. Then we learn that beyond addition, multiplication, and exponentiation there's a never-ending hierarchy of hyper-operations, starting with tetration (repeated exponentiation).

The trouble is that, up to this point, the ways we've used to try to represent this hierarchy are the equivalent of old technology, which hampers any attempts at progress.

We've said that, just as we can denote an exponential as a^n, so a tetration can be represented by ^na. So far, so good – but what comes next? The trick of moving the n to the other side of the base number to indicate tetration can't be repeated for the next level of hyperoperator or the one after that. Some more powerful and flexible scheme is needed to notate numbers that are far larger than any we've encountered up till now.

In fact, various such schemes have been devised over the past century or so. Some of these have been adopted by mainstream mathematicians and find their way into academic papers. Others are better known on the fringes of the subject and are most commonly used in popular-level writings about big numbers (such as this book!) or the webpages of googologists – mathematicians, both amateur and professional, who, in their spare time, study, name, and find new ways to represent large numbers.

We've seen that the mathematical operations with which we're most familiar – addition, multiplication, and exponentiation – are part of a never-ending sequence called the hyperoperation sequence. Strictly speaking, this begins with the successor operation. Then it continues, after addition, multiplication, and exponentiation, with tetration, pentation, and so on, indefinitely. Each successively more powerful operator amounts to merely the repeated action of the operator below it in the hierarchy. A conventional way of representing specific operations in this sequence uses H notation. For instance, multiplying 5 by 4 would be shown as $H_2(5,4)$. This indicates that we're dealing with the second hyperoperator (multiplication) after the zeroth one (succession), H_0, and the first (addition), H_1. Or take another example: $H_3(2,5)$. The third hyperoperator is exponentiation, so $H_3(2,5)$ means $2^5 = 2 \times 2 \times 2 \times 2 \times 2 = 32$. After this comes H_4, which is tetration, or repeated exponentiation. $H_4(4,3)$, for instance, is 4 to the power 4 to the power $4 = 4^{256} = 1.34 \times 10^{154}$ (approximately).

A different but equivalent system of notation uses squared brackets to indicate the level of hyperoperator and takes the form $a[n]b$. Using the examples we've just considered:

$$H_2(5,4) = 5[2]4,$$
$$H_3(2,5) = 2[3]5, \text{ and}$$
$$H_4(4,3) = 4[4]3$$

In 1976, American computer scientist and mathematician Donald Knuth proposed another and, again, exactly equivalent, way of writing numbers in the hyperoperator sequence. It's known as up-arrow notation. Knuth had already established a reputation as a leader in the field of theoretical computer science with his multi-volume work *The Art of Computer Programming*, the first part of which came out in 1968. He was also the first to publish a book on a new, expansive system of numbers, proposed by English mathematician John Conway, which includes not only all finite numbers but also all infinitely large and infinitely small ones as well.

During a lunchtime meeting with Knuth in 1972, Conway had explained the ideas behind his extraordinary new number system. Knuth was immediately captivated. The following year, spent at the University of Oslo, during a week-long break from his work on *The Art of Computer Programming*, Knuth penned an introduction to Conway's mind-bogglingly large collection of numbers, which Knuth christened 'surreal numbers'. 'I believe it is the only time', wrote Knuth, 'that a major mathematical discovery has been published first in a work of fiction.' Knuth's novelette, titled *Surreal Numbers: How Two Ex-Students Turned on to Pure Mathematics and Found Total Happiness*, takes the form of a dialogue between a young couple who studied maths together at university and are now stranded on an island. The two come across stone tablets on the beach, which give them clues about how to construct Conway's number system, starting from the empty set (the set that has no 'elements' or members) and a few simple rules. As Knuth pointed out, his main goal in writing the book, beyond explaining surreal numbers, was to give a feel for what research in mathematics is like. The two lovers don't follow a straight line in their path to understanding the surreals but, as in the real world of research, make mistakes and have to retrace some of their steps before finding a better way forward.

Knuth continued his adventures into deep dark number-land by announcing up-arrow notation in a paper that appeared in the journal *Science*. A single up-arrow represents exponentiation. For example, $2 \uparrow 4 = 2 \times 2 \times 2 \times 2 = 2^4 = 16$.

Two up-arrows take us to the next rung on the ladder of hyperoperators: tetration. For example,

$$2 \uparrow\uparrow 4 = 2 \uparrow (2 \uparrow (2 \uparrow 2))$$
$$= 2 \wedge 2 \wedge 2 \wedge 2 = 2^{16} = 65,536$$

Adding another up-arrow brings us to the level of pentation:

$$2 \uparrow\uparrow\uparrow 4 = 2 \uparrow\uparrow (2 \uparrow\uparrow (2 \uparrow\uparrow 2))$$

Now we continue, evaluating from right to left:

$$
\begin{aligned}
2 \uparrow\uparrow (2 \uparrow\uparrow (2 \uparrow\uparrow 2)) &= 2 \uparrow\uparrow (2 \uparrow\uparrow (2 \uparrow 2)) \\
&= 2 \uparrow\uparrow (2 \uparrow\uparrow 4) \\
&= 2 \uparrow\uparrow 65{,}536
\end{aligned}
$$

Evaluating this number gives a power tower of 2s of height 65,536. Even a tower of height 5 produces a ridiculously big number, with approximately 26,300 digits. Adding another 2 to the tower utterly dwarfs the googolplex. Yet there are still 65,530 more 2s left to go!

How about the hexation $2 \uparrow\uparrow\uparrow\uparrow 4$? This turns out to be:

$$2 \uparrow\uparrow\uparrow (2 \uparrow\uparrow\uparrow (2 \uparrow\uparrow\uparrow 2))) = 2 \uparrow\uparrow\uparrow (2 \uparrow\uparrow\uparrow 4)$$

so you would need to take 2 pentated to this mighty power tower of height 65,536. Starting with 2, this would involve making a series of power towers, with the result of each being the height of the next. First we have 2, then $2 \uparrow\uparrow 2$ = 4, then $2 \uparrow\uparrow 4$ = 65,536, then $2 \uparrow\uparrow 65{,}536$, which is the previous power tower, and so on, with the number of such power towers being itself a power tower of height 65,536.

Like the googolplex, which it makes appear ludicrously small, $2 \uparrow\uparrow\uparrow\uparrow 4$ *couldn't* be written out in full or held in any computer memory that could actually be constructed, present or future. It's simply a number vast beyond the capacity of the universe to display in decimal digit form.

Yet we *can* represent it, very succinctly, in the form of those six symbols: 2 ↑↑↑↑ 4.

The result of applying up-arrows depends strongly not only on the number of up-arrows but also on the numbers upon which the up-arrows operate. In going from 2 ↑ 4 to 2 ↑↑ 4, we went from 16 to 65,536. But let's say we change the 2 to a 5 and the 4 to a 3. In the case of a single up-arrow, $5 \uparrow 3 = 5^3 = 5 \times 5 \times 5 = 125$. But the tetration 5 ↑↑ 3 is $5 \uparrow (5 \uparrow 5) \doteq 5 \uparrow 5^5 = 5^{3,125}$. The value of five multiplied by itself 3125 times is approximately $1.911 \times 10^{2,184}$. Remember, the earlier tetration we looked at, 2 ↑↑ 4, came out to be just 65,536, so, clearly, the size of the numbers being tetrated has a dramatic effect on the rate of growth of the outcome. Imagine how much more spectacular will be the explosion in values when we move to pentation, represented by three up-arrows.

The mesmerising scale of tetrated and pentated numbers becomes even more evident when we bear in mind that numbers such as 2, 3, 4, and 5 are small. Imagine the result of the tetration 53 ↑↑ 19 or the massively greater still pentation 74 ↑↑↑ 136. Yet even these numerical titans are dwarfed by the outcomes if we move to hyperoperators associated with even more up-arrows.

You may be thinking, why on Earth do we need ways to represent numbers that are so monstrously huge? Of course, there's no *practical* reason at all. There's nothing in the real, physical universe that corresponds with anything anywhere near as big as the number we've just written out in full, never mind all the decimal digits of 2 ↑↑↑ 3 when it's been completely unpacked. But we're not exploring the limits of the material world here. We're not confining ourselves to what's meaningful or useful from a practical standpoint. We're in the realm of pure mathematics, where the 'space' available for containing and representing numbers, in any way we choose, is unlimited and where there's no need to ask if there's any point to something just because it doesn't have a physical correlate.

Knuth's up-arrow notation is impressively powerful compared to anything else we've come across so far. But you can get just as tired of writing long lines of up-arrows as you can of writing a lengthy string of zeros. Here, for example, is a perfectly valid number specified in up-arrow notation:

$$8 \uparrow 17$$

The mind freezes at what this might be like if it were possible to write it out as an ordinary number. But even shown in up-arrow form, it looks unwieldy. How many up-arrows are there in that row? As it happens, there are 39. But the only way you could discover that for yourself is to count them, which is a pain. Far better to write it in the more compact form $8 \uparrow^{39} 17$, so that we know at a glance what we're dealing with.

It's easy to see that Knuth's up-arrow notation is equivalent to the two shorthand schemes we mentioned earlier for representing numbers in the hyperoperator sequence. The only difference is that up-arrows start to be used only at the stage of exponentiation, so the number of up-arrows is always two less than the H subscript or the number in the squared brackets of the other methods. Using the examples we looked at when describing the H operator and squared-bracket schemes:

$H_2(5,4) = 5[2]4 = 5 \times 4$ (no up-arrows for multiplication)
$H_3(2,5) = 2[3]5 = 2 \uparrow 5$ and
$H_4(4,3) = 4[4]3 = 4 \uparrow\uparrow 3$

Each of these equivalent schemes involves what's called a *binary* or *dyadic* operator because it acts on just two elements – the numbers in the brackets after the H operator, on either side of the squared brackets, or on either side of the up-arrows. As for the three different parameters – the level of the hyperoperator and the two elements acted upon – these can be referred to as the rank, base, and exponent (or hyperexponent), respectively. So, for instance, in $H_3(2,5)$, 3 is the rank, 2 is the base, and 5 is the exponent. In the case of up-arrows, we need to add 2 onto the number of up-arrows to get the rank. For example, if we have $6 \uparrow\uparrow\uparrow 3$, the rank is 5 (pentation), the base is 6, and the exponent is 3, and we could say $6 \uparrow\uparrow\uparrow 3$ as 'the third pentation of 6'.

We can represent some terrifyingly big numbers in these three equivalent forms of hyperoperator notation. Numbers like $H_{43}(15,7)$, 24[73]18, and $62 \uparrow^{101} 29$ are spectacularly vast. But take up-arrow notation: what if we keep adding more up-arrows – many, many more? For example, there's such a number as

$$43 \uparrow^{26,000,000,000,000,000,000,000,000,000,000,000,000} 85$$

with 26 million trillion trillion trillion up-arrows between the 43 and the 85.

Now the problem becomes keeping track not of the number of up-arrows but of the number of zeros telling us the number of up-arrows! We could collapse the exponent after the up-arrows to 26×10^{42}. But then we're getting back into the inconvenience, and typographic nightmare, of having to write exponents of exponents, and then of exponents of exponents of exponents, and so on, which returns us to the same situation we had before and certainly wins you no friends among your publisher's production team. The fact is that even Knuth's up-arrow notation eventually reaches a stage where it's impracticable.

You won't be surprised to hear that other ingenious schemes have been devised for representing fantastically big numbers. One of these is named Steinhaus–Moser notation after its creators, Hugo Steinhaus and Leo Moser. Steinhaus was a Jewish-Polish mathematician who earned his PhD in 1911 at Göttingen University under the supervision of David Hilbert, the greatest mathematician of his day. Together with his compatriot, Stefan Banach, Steinhaus went on to discover one of the most important results in a branch of maths known as functional analysis. He's also considered to have been a leading light in the development of both game theory and probability theory. In his popular-level book *Mathematical Snapshots*, published in 1950, Steinhaus suggested a way of representing large numbers using triangles, squares, and circles. He starts by defining a number, n, inside a triangle to mean n^n. For example,

$$\triangle\!\!\!2 = 2^2 = 4$$

Next, n written inside a square denotes n inside n triangles. So, for example,

$$\boxed{2} = \triangle\!\!2\!\!\triangle = \triangle\!\!2^2\!\!\triangle = 4^4 = 256$$

Finally, n inside a circle denotes n inside n squares. Continuing with our example, in which $n = 2$, we arrive at the result that:

$$\bigcirc\!\!2 = \boxed{\boxed{2}} = \boxed{\triangle\!\!2\!\!\triangle} = \boxed{\triangle\!\!2^2\!\!\triangle} = \boxed{4^4} = \boxed{256}$$

Steinhaus called this number – 256 inside a square – the mega. The question is: just how big is it?

The mega is the same as 256 written inside 256 nested triangles. The innermost triangle means 256^{256}, which in itself, written out in full, is a pretty impressive number:

32,317,006,071,311,007,300,714,876,688,669,951,
960,444,102,669,715,484,032,130,345,427,524,655,
138,867,890,893,197,201,411,522,913,463,688,717,
960,921,898,019,494,119,559,150,490,921,095,088,
152,386,448,283,120,630,877,367,300,996,091,750,
197,750,389,652,106,796,057,638,384,067,568,276,
792,218,642,619,756,161,838,094,338,476,170,470,
581,645,852,036,305,042,887,575,891,541,065,808,
607,552,399,123,930,385,521,914,333,389,668,342,
420,684,974,786,564,569,494,856,176,035,326,322,
058,077,805,659,331,026,192,708,460,314,150,258,
592,864,177,116,725,943,603,718,461,857,357,598,
351,152,301,645,904,403,697,613,233,287,231,227,
125,684,710,820,209,725,157,101,726,931,323,469,
678,542,580,656,697,935,045,997,268,352,998,638,

215,525,166,389,437,335,543,602,135,433,229,604,
645,318,478,604,952,148,193,555,853,611,059,596,
230,656

Let's call this number $n1$. It has 617 decimal digits and written in terms of powers of ten is approximately 3.23×10^{616}. It's obviously a heck of a lot bigger than a googol, which is a mere 10^{100}. But we've only just started to work our way up through the set of triangles within triangles in order to evaluate the mega. Next we have to figure out the value of $n1^{n1}$: a 617-digit-long number raised to the power of itself. That's not possible to compute because the answer has about 10^{616} digits – a number immensely larger than the number of subatomic particles in the universe. And beyond this, we have 254 triangles to go!

The mega, which appears so simple and compact when written as just a 2 inside a circle, is a monster in disguise. It makes the googolplex and Skewes' number look almost indistinguishable from zero. We can't know the exact value of the mega but, strange as it may seem, it isn't hard, given a few hours' computing time on a laptop, to figure out its final few digits. We'll never be able to work out the first digit of the mega, or the vast bulk of those that follow, but the last of them are known, with absolute certainty, to be 42656. The 6 at the very end, for instance, is easy to explain since it comes from multiplying 6 by itself many times in the computation of an immense power tower of 256s, and any whole number power of 6 ends in 6.

It's not hard to show (although we won't do it here) that, in terms of up-arrows, the value of the mega lies somewhere between 10 ↑↑ 257 and 10 ↑↑ 258. That may seem surprisingly modest. After all, we've been giving the mega a big build-up and illustrating how mind-blowingly huge it is when we work our way up through the different levels of its notation in Steinhaus form. Yet, its upper and lower bounds can be written quite compactly using up-arrows. The mega *is* truly colossal by the standards of the ordinary numbers with which we're used to dealing. The fact that 10 ↑↑ 257 < mega < 10 ↑↑ 258 just goes to show how powerful is the operation of tetration, compared with the more familiar, lower-order operations of addition, multiplication, and even exponentiation, and also how potent are Knuth's up-arrows as a way of representing large numbers.

In *Mathematical Snapshots*, Steinhaus named an even bigger number, the megiston, defined as

and which, in terms of up-arrows, can be expressed as a pentation with a value between 10 ↑↑↑ 11 and 10 ↑↑↑ 12.

Austrian-Canadian mathematician Leo Moser extended the notation devised by Steinhaus to include pentagons, hexagons, heptagons, and, in general, any polygon with *x* number of sides and any number written inside it. The Steinhaus–Moser system replaces the circle with a pentagon and then uses polygons with more than five sides to enable a startlingly rapid growth in the size of numbers that can be represented. In general, a number *n* written inside an *x*-sided polygon is the equivalent of *n* inside *n* nested polygons each with *x* − 1 sides.

The number known as the moser is defined as 2 written inside a megagon. Here we're using 'megagon' to mean a polygon with a mega number of sides, as distinct from its other meaning, in conventional plane geometry, which is a polygon with a million sides. Remember that the mega is itself a colossal number – somewhere between $10 \uparrow\uparrow 257$ and $10 \uparrow\uparrow 258$ – and that we're now dealing with a mega-sided polygon. So it's evident that the moser is very, very big indeed. In terms of up-arrows the value of the moser lies between $2 \uparrow^{mega-2} 3$ and $2 \uparrow^{mega-2} 4$. That's breathtakingly large when you consider the awesome growth in size that comes with the addition of each and every up-arrow and that here we're talking about a *mega* number of up-arrows (minus 2). Whereas the mega is a tetration-level number (with a value between $10 \uparrow\uparrow 257$ and $10 \uparrow\uparrow 258$), the moser is so huge it involves a tetration-level number of up-arrows!

With the moser we've arrived at something new: a number that, expressed in terms of up-arrows, involves a *recursion* of the number of up-arrows. In art, music, language, computing, and maths, recursion pops up in all kinds of different guises, but always it refers to something that feeds back into itself. In some cases this just leads to an endless, repetitive loop. For example, there's the joke glossary entry: 'Recursion. See *Recursion*.' On a more elaborate scale, a recursive loop appears in Maurits Escher's *Print Gallery* (1956), which shows a gallery in a city in which there's a picture of a gallery in a city in which… In engineering, a classic example of recursion is feedback, where the output from a system gets routed back as input. It's a familiar problem for performers, such as rock musicians, on stage, and often happens if a microphone is located in front of a loudspeaker to which it's connected. Sounds picked up by the mic come out of the speaker, having been amplified, then re-enter the mic at a higher volume to be amplified again, and so it goes on until, very quickly, the familiar, ear-piercing squeal of feedback emerges. Recursion in maths works along similar lines. A function takes the place of an electronic system, such as a microphone-amplifier-loudspeaker combination, and calls upon itself, so that it feeds its own output back as input.

The well-known Fibonacci series is a simple case of mathematical recursion at work. The series starts with the numbers 1, 1. These are then added to give 2, and the series continues by adding the previous two numbers together, to produce the Fibonacci numbers 1, 1, 2, 3, 5, 8, 13, 21, 34, 55, ... In the seventeenth century, German astronomer and mathematician Johannes Kepler showed that the ratio of successive terms in the Fibonacci series (1/1, 2/1, 3/2, 5/3, 8/5, ...) approaches a number that (like pi) crops up a great deal in maths, known as the golden ratio, ϕ (phi), which is approximately 1.618. In 1765, Swiss mathematician Leonhard Euler published a formula, now called (through misattribution) the Binet formula, showing that the Fibonacci numbers grow at an exponential rate equal to ϕ.

Shapes known as fractals are produced by recursion. One of the defining features of a fractal is self-similarity, which means that the whole object has the same, or a similar, shape to one or more of its parts. A simple example is the Sierpiński triangle. It has the overall shape of an equilateral triangle and is constructed by repeatedly removing triangular subsets, as shown in Figure 4.2.

FIGURE 4.1: The Sierpiński triangle is a fractal shape that evolves by recursively applying a simple set of rules.

FIGURE 4.2: The Sierpiński triangle after eight iterations.

From now on, we're going to find that recursion, in one form or another, is at the heart of most of our attempts to represent and define ever-larger numbers. In the case of the moser, we see a number (the mega), described using up-arrows, being used as the starting point for a recursive process that ends up specifying the number of up-arrows needed to describe the moser. There's nothing to stop us repeating the process and defining a new number – call it the 'monster-moser' – in terms of a moser number of up-arrows. Let's define the monster-moser to be $2 \uparrow^{moser} 3$. And why stop there? Somewhere in the infinite space of the mathematical cosmos is a number we'll dub the 'mega-monster-moser' equal to $2 \uparrow^{monster-moser} 3$. And so on and so on. The point is, the way the moser is defined gives us an early glimpse of the power of recursion and what might lie far beyond the land of Knuth's up-arrows.

Of course, Steinhaus–Moser notation, as a pictorial form of large-number representation, is impossible to use in practice, with the exception of a few simple cases like the ones we've mentioned. Even a polygon with a few dozen sides is hard to draw. One with a mega number of sides (needed to show a moser) is a physical impossibility: it would be indistinguishable from a true circle even at the subatomic level. Yet, despite its impracticality as a kind of mathematical shorthand, Steinhaus–Moser notation is important in revealing how potent recursion can be. In fact, it's one of the earliest systems to be devised that can take us into the same territory as something called a *fast-growing hierarchy*. We'll have much more so about this type of hierarchy later because it's the most widely accepted and proven method of reaching some of the largest numbers that can be well defined in all of mathematics. But we need to approach it gradually, step by small step, as we'll be doing in the next few chapters, in order to understand it properly.

Already, in our travels, we've come across numbers that are way too big to grasp, in their entirety, with our brains. A billion, or even a trillion, we can dimly imagine. But something as huge as the mega, never mind the moser, dwarfs our ability to comprehend it. Anything that's outside our ability to sense or experience directly poses an intellectual challenge. We face similar obstacles in trying to grasp the goings-on of the quantum world or the scale of the cosmos. The best we can hope to do is follow and understand the steps by which numbers such as the mega and moser are clearly and unambiguously defined.

This raises an interesting philosophical point – one that we'll come back to in more detail later. We said earlier that we can prove, beyond a shadow of doubt, that the mega ends with '42656' but that we can never know how it starts or what almost all of the digits are that follow. We can say that the mega exists, in some sense, because we can define it exactly. It's equal to 256 raised to a series of powers of 256 – a power tower – that's 256 levels high. This description is enough to pin down a specific number. But, here's the problem: if the number so described can never be written out in full (in decimal form) and if the overwhelming majority of its digits are unknown – and, in practice, unknowable – what's the status of its existence? If it's impossible, in the real world we inhabit, ever to discover what the first digit of mega is, what can we say about its nature? Is that first digit – a number between 1 and 9 – like most of the rest of mega, in some probabilistic state akin to the uncertain position of an electron in an atom? That hardly seems reasonable: maths isn't governed by the rules of quantum mechanics. Is its existence confined to some Platonic universe of mathematics outside of the physical realm? To put the question more generally: whereabouts are mathematical objects before we've established, with our minds or machines, their exact nature or value?

The Greek philosopher Plato maintained that abstract objects can exist independently, of their own accord. In other words, says Platonism, there's a third realm distinct from the outer world we can sense and the inner world of consciousness. Most mathematicians go about their business content with this idea of a Platonic realm, inhabited by such things as numbers. But in the final analysis, when we try to understand deeply the interplay between maths, mind and matter, we have to confront the issue head on.

Take the moser, for instance. It can be pinned down in the cleverly compact form of Steinhaus–Moser notation and there's an easy method *in principle* for figuring it out.

But its exact value will never be known because the universe lacks – and will always lack – the resources, in terms of time, space, matter, and energy, to compute it. What does it mean, then, to say that, for example, the first digit of the moser exists? *Where* does it exist and in what form?

Exactly the same question could be applied to the decimal expansion of the number pi. We know precisely what pi is. Its definition, in geometric form, is much easier to understand than the shapes-within-shapes definition of the moser. Pi is just the ratio of the circumference to the diameter of a circle. It starts off 3.14159... and then carries on forever in decimal form, never going into an endlessly repeating pattern. Its value has been calculated by computer to about 10 trillion decimal places. The digits we know have somehow been brought into the physical universe, their value established for all time. Because of this – because these digits are essentially frozen and unchanging – they can't be said to form a random sequence. Randomness implies uncertainty, and there's nothing uncertain, for example, about the five-hundredth decimal place of pi. Are these digits in some sense more real than those we've yet to calculate?

Think about the value of the next decimal place beyond all those that have been computed so far. We know that the next such number exists because we can prove that pi is irrational and therefore must have an infinitely long, non-repeating decimal expansion. But, at the moment, we can't say which of the digits, 0 to 9, the next unknown digit is. We can't even say if it's slightly more likely to be one digit, say a 2, than any other because we don't know whether or not pi is 'normal'. A number is said to be normal if no digit occurs more frequently than any other. So, if a number in base 10 (in other words, expressed as a decimal) were normal, every digit from 0 to 9, over a long enough sequence, would crop up with equal likelihood.

In the future, we'll doubtless be able to add tremendously to the accuracy with which pi is known. But there'll always be a next decimal place that we haven't yet figured out. No matter how fast our computers become or however long they crunch away trying to calculate pi ever more precisely, we'll never have a clue as to what the next digit will be.

Someone who holds a Platonic, or realist, view of mathematics would argue that all mathematical objects, including numbers such as pi and the moser, exist in their entirety, independently of our knowledge of them. This tends to be the default position of professionals in the field: that all of mathematics is somehow out there waiting, like buried treasure, to be uncovered. Quite where 'out there' is remains a debatable point. Set against this standard philosophical view are various forms of anti-realism, which question whether maths has any independent existence outside of the mind or the physical phenomena that it happens to describe.

The truth is, most working mathematicians, like the majority of scientists, don't go in much for philosophy, at least not as part of their day job, because it rarely has any bearing on what they're actually doing. In the same way, most mathematicians have no need to think about gigantic numbers, or ingenious methods of writing them. They're happy to acknowledge that numbers go on forever and that somehow, somewhere, fantastically large numbers exist. If they do philosophise or think about curious topics outside of their speciality, it tends to be in their off-duty hours, when they allow their minds to roam free.

By their nature, mathematicians and scientists tend to have a playful side. Unsolved problems, puzzles, paradoxes, and other stuff that fires the imagination are often what drew them to their subject to begin with. Scientists get hooked as kids by science fiction and by popular-level books on space, subatomic physics, and the like. In the case of the authors, David was inspired when young by the likes of Verne, Wells, Clarke and Asimov, whereas Agnijo's desire to become a mathematician was kindled especially by the writings of Ian Stewart. Many mathematicians were weaned on the writings of Martin Gardner in *Scientific American*, classics such as Kasner and Newman's 1940 *Mathematics and the Imagination*, and more recent books, like Douglas Hofstadter's *Gödel, Escher, Bach: An Eternal Golden Braid*, which highlight exotic or recreational aspects of a subject that to others may appear dry and dull.

Very often new ideas about large numbers are first made public in writings aimed at a broad audience. 'Googol' and 'googolplex' – names invented by a child – entered the popular lexicon by way of Kasner and Newman's 1940 classic. Steinhaus first described his circle notation in the Polish edition of *Mathematical Snapshots*, published in 1937. Martin Gardner, in his monthly 'Mathematical Games' column, introduced the world to other ways of representing mind-bogglingly big numbers, which might otherwise have remained in obscurity.

Some outsized numbers, and ingenious schemes for reaching them, were devised by mathematicians in their spare time, almost as an amusement. Others arose for a definite reason: the effort to solve a particular problem in maths. This was the case with Skewes' number, which we met in the last chapter. At the time of its announcement, in 1933, it was the biggest number ever put forward in the context of serious mathematical research. It held that title for about forty years until it was supplanted by a number so large that it will take us a whole chapter to fully appreciate.

CHAPTER 5

G Whizz

IT'S PRETTY UNUSUAL to come across someone who's served as president of both the American Mathematical Society and the International Jugglers' Association. Ronald Graham, who died in 2020, was a man of diverse talents. But the reason he became known outside his professional circles was really down to a single number, named after him, which earned him an entry in *The Guinness Book of Records* and a feature in *Ripley's Believe It or Not!* 'Graham's number' is one of the largest numbers ever to be used in a mathematical proof. It's a number so large that it renders utterly insignificant every other number that we've encountered so far.

Ron Graham was born in 1935, in Taft, California, where his father worked in the nearby oil fields. The family were always on the move when Graham was young, mostly between California and Georgia, as his dad shifted from one oil or shipbuilding job to another. Being exceptionally bright, an effect of this itinerant lifestyle was that, when starting a new school, Graham would often be put in a class of older children so that he skipped grades and made rapid academic progress.

FIGURE 5.1: Ronald Graham in 1998.

Astronomy was his first passion but this was soon replaced by the subject at which he excelled: mathematics. At the age of fifteen he won a scholarship to the University of Chicago and began his studies there without ever graduating from high school. Strangely, in his three years at Chicago he didn't take a single maths course due to the fact that his scholarship score in this area was so high. On the upside, he did learn gymnastics and developed a special talent for juggling and trampoline.

In 1954, Graham moved to the University of California at Berkeley for a year, where he majored in electrical engineering. He also took a course in number theory, given by Derrick

Henry Lehmer, which shaped his future career. To help support himself while studying, he and two other students performed as a trampoline troupe at schools, supermarket openings, and even the circus. In 1962, he earned his PhD at Berkeley, under Lehmer's supervision, with a thesis titled 'On Finite Sums of Rational Numbers'.

At a conference in Boulder, Colorado, in 1963, Graham first met Hungarian number theorist Paul Erdős (pronounced 'airdosh') – one of the most prolific mathematicians of the last century. Erdős spent almost every waking hour solving mathematical problems while essentially living out of a battered suitcase, which he took with him from one conference or university to another. Colleagues opened their homes to him, more than happy to exchange board and lodging for deep mathematical conversation, and he donated much of his income to good causes or towards prizes for correct solutions to problems he suggested. Graham and Erdős had similar interests, became close friends despite a generational difference in age, and wrote nearly 30 papers together. In total, Erdős published around 1,500 papers – more than any other mathematician in history – many of them with collaborators. Graham popularised the idea of the 'Erdős number'. Anyone who had co-authored a paper with Erdős had an Erdős number of 1. Someone who'd co-authored a paper with someone who'd co-authored a paper with Erdős had an Erdős number of 2, and so on.

One of the subjects that fascinated both Graham and Erdős was Ramsey theory. This was a relatively new field that had been pioneered by English mathematician and philosopher Frank Ramsey, brother of Michael Ramsey, who became Archbishop of Canterbury. Frank Ramsey graduated in 1923 as Senior Wrangler (top of his class) in mathematics at Trinity College, Cambridge, where one of the authors (Agnijo) now studies. At the age of nineteen he learned German in less than a year and then produced the first English translation of Wittgenstein's monumental work *Tractatus Logico-Philosophicus*, the goal of which is to explore the relationship between language and reality and define the limits of science. Subsequently, Ramsey was instrumental in persuading Wittgenstein to resume his philosophical research and, at the same time, come back to Cambridge where he'd previously taught.

In 1928, aged twenty-five – a little over a year before he died following an abdominal operation – Ramsey wrote a paper called 'On a Problem of Formal Logic'. The main aim of the paper was to solve an aspect of what mathematicians called the *Entscheidungsproblem* (German for 'decision problem'). Proposed by David Hilbert and fellow German mathematician Wilhelm Ackermann, this asks whether a step-by-step method always exists for deciding whether a given statement can be proved from basic starting assumptions, or axioms, using the rules of logic. Along the way to the main conclusion of his paper, Ramsey derived what, at the time, seemed like a fairly minor result but which turned out to be of fundamental importance. It's now known as Ramsey's theorem. The simplest illustration of it is called the theorem of friends and strangers. Suppose there are six people at a party. If you consider any two of them, either they've met before, in which case we'll call them 'friends', or they've never met before and so are strangers.

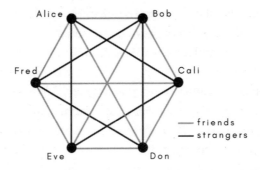

FIGURE 5.2: A possible set of connections between six individuals, each pair of which may be either friends or strangers.

Figure 5.2 shows an example. Alice has met Bob before so that they're friends. However, Alice and Cali haven't previously met and so are strangers. The theorem of friends and strangers states that, among six partygoers, either at least three will be mutual strangers (none of them having met the other two before) or at least three will be mutual friends. In the case illustrated here, Alice, Bob, and Don, for example, are mutual friends because each one of them has met the other two – in other words, they're pairwise mutual friends. No matter how you draw the connections, you'll always find at least one such triangle of pairwise mutual friends or pairwise mutual strangers.

Ramsey's theorem in its original form, you won't be astonished to learn, isn't framed in terms of people at parties. Instead it focuses its attention on the more abstract and general subject of graph theory. Everyone's familiar with the type of graph that shows how one quantity varies with another: curved or straight line plots of x against y that we learn about in school. However, graph *theory* deals with graphs of a completely different kind. Graph theory belongs to a branch of maths known as discrete mathematics, and the objects it studies don't vary smoothly but instead take on distinct, discrete values. The graphs in graph theory consist of points, also known as nodes or *vertices*, any one of which may be connected to one or more other vertices by lines, or *edges*.

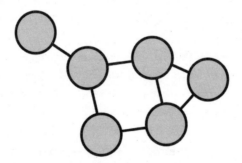

FIGURE 5.3: An undirected graph consisting of six vertices and seven edges.

These graphs come in different types: directed (those with arrows on the edges), undirected, complete, finite, weighted, and so on. A complete graph is one in which every pair of vertices is connected by a unique edge. Graphs may also be 'labelled', for example by colouring the vertices or edges. Another important concept in graph theory is the 'clique'. A clique is a subset of vertices distinguished by the fact that every two distinct vertices in a particular clique are adjacent.

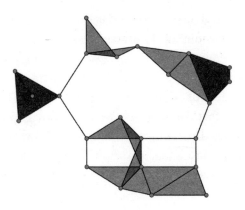

FIGURE 5.4: A graph with 23 one-vertex cliques (the vertices themselves), 42 two-vertex cliques (the edges), 19 three-vertex cliques (light and dark shaded triangles), and 2 four-vertex cliques (dark shaded areas).

Ramsey's theorem has to do specifically with complete graphs that have coloured edges. It says that in any complete graph with coloured edges it's always possible to find cliques of the same colour providing the graph is large enough. That may seem like a pretty obscure piece of information. But it turned out to be the seed from which sprouted an entirely new branch of mathematics – Ramsey theory. The overarching idea in Ramsey theory is that, however disordered a system may appear to be, provided that it's sufficiently large, it must contain some order. A typical problem in it runs along the lines: how many elements of a structure are needed to guarantee that a particular property will hold?

In 1971, Ron Graham and a colleague, Bruce Rothschild, published a paper on Ramsey theory that had to do with cubes, and not just the common or garden variety but cubes in any number of dimensions – 'hypercubes' that might have a dimensionality of four, five, or more. Everyday cubes, like sugar cubes or dice, are 3D shapes and have 2^3, $= 8$, corners or vertices. A 4D cube, sometimes referred to as a tesseract, can't be properly visualised (because we think only in three dimensions) but can be shown mathematically to have 2^4, $= 16$, vertices. A 5D hypercube has 2^5, $= 32$, vertices and, in general, an n-dimensional cube has 2^n vertices.

Think about an ordinary (3D) cube and connect every vertex to every other vertex by a line. There'll be 28 lines in all, as shown in Figure 5.5.

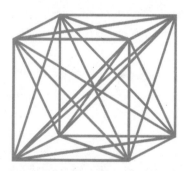

FIGURE 5.5: A cube in which every vertex is connected to every other by a line.

Now let's colour each of the lines red or blue in any way we like. In the illustrations here, dark grey represents blue and light grey represents red. One of the many possibilities we might end up with is shown in Figure 5.6.

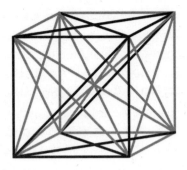

FIGURE 5.6: A version of Figure 5.5 in which some lines have been coloured light grey and others dark grey.

Now, here's the question. No matter how we decide to colour the lines, is it always possible to find four vertices *in the same plane* so that all the lines connecting them are the same colour? In the graph of the cube we've just coloured, there are indeed four such vertices, as shown in Figure 5.7.

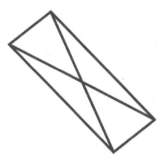

FIGURE 5.7: An example of four vertices from Figure 5.6, which lie in the same plane and are all connected by lines of the same colour (light grey).

But if, instead, we'd decided to colour the bottom edge in Figure 5.6 blue (dark grey), there wouldn't have been any set of four vertices that satisfied our criterion. Having found a counterexample, we know that, in three dimensions at least, cubes can't meet the conditions of our problem.

In their 1971 paper, Graham and Rothschild asked a generalised form of the question we've just talked about. Connect each and every pair of vertices of an n-dimensional hypercube, they said, to make a complete graph having 2^n vertices. Colour each of the edges red or blue. Is there any dimensionality of hypercube so that among all the possible colourings at least one *must* give rise to a complete subgraph of four coplanar vertices? (A complete subgraph is just a complete graph that forms part of a larger complete graph, as in the case of Figure 5.7.) The answer, they found, is 'yes'. Provided the dimension n is large enough, it's guaranteed that, no matter how the vertices of the hypercube are coloured, there'll always be a slice or subgraph that has a single colour.

Graham and Rothschild weren't able to determine the smallest value of n that would force the conditions of their problem to be met. In fact, no one's yet done that. However, they were able to show that the *minimum* value it could be was 6 and the *maximum* value – the upper bound – was a stupendously large number, which, in a simplified form, has come to be known as Graham's number.

You might think that's a pretty major uncertainty – knowing only that the value of something is between 6 and a number that's inconceivably vast – and you'd be right. But the wide uncertainty isn't due to the incompetence of the mathematicians involved but instead to the incredible complexity of many problems in Ramsey theory. We needn't look any further than the friends-and-strangers-at-a-party scenario, considered earlier, to get a sense of the difficulties faced. Remember we said that you need a group of six people to ensure that there'll be three friends or three strangers. In the notation of Ramsey theory, this can be written as $R(3,3) = 6$, where $R(3,3)$ is called a Ramsey number. In general, the Ramsey number $R(m, n)$ gives the minimum number of guests who must be invited to a party to guarantee that at least m will know each other or at least n will be strangers. It's known that $R(4,4) = 18$. But beyond that, things get a bit hazy. The trouble is that the number of possible combinations in Ramsay theory grows explosively fast with the size of the problem. A graph representing four people and the possible pairwise connections between them has just six edges and 2^6, $= 64$, possible red/blue (friend/stranger) colourings. But the graph for six people has fifteen edges and 2^{15}, $= 32,768$, ways of being coloured.

Mathematicians have yet to figure out the value of $R(5,5)$, in other words the size of party you'd need to ensure that at least five people will be acquainted or five will be strangers. The answer's known to be between 43 and 49, and is strongly suspected of being 43, but no one's yet been able to prove it. The 'brute-force' method of checking all the possible ways of colouring a graph of 43 people with two different colours is a non-starter. Such a graph has 903 edges and therefore 2^{903} possible ways of being labelled using two colours – a number far exceeding the number of subatomic particles in the observable universe and the capacity of any computer to check case by case.

The problem investigated by Graham and Rothschild is similar in nature to that of the friends and strangers situations we've looked at but involves a lot more vertices and edges. Any possibility of checking individually all the different ways of colouring the connections between vertices of multidimensional hypercubes is out of the question. Instead, mathematicians have to seek other, subtler methods of proving if and when a particular condition is met. As in the case of Graham's number, these often succeed only in establishing bounds for the solution of a problem rather than the solution itself. Graham's number, as we're going to find out, is so spectacularly large that it doesn't seem like a very useful upper bound, especially when the actual answer to the problem it addresses is suspected of being almost comically small – possibly in the teens! But, the fact is, it's a start. The same was true of Skewes' number. It, too, was stated as an upper bound to a problem and is a large number – $10^{10^{10^{34}}}$ – although not in the same league as Graham's number. It served as a starting point for other mathematicians to try to narrow the gap between the known upper bound and the actual solution.

More recently, progress involving upper bounds has been made on a big open question that was first proposed in 1846 by French mathematician Alphonse de Polignac (though, doubtless, others had thought about it before). The twin primes conjecture, or Polignac conjecture, asserts that there are infinitely many pairs of prime numbers that differ by only two, such as 11 and 13, 29 and 31, and 101 and 103. It's a claim that's easy to understand and almost universally believed to be true but is stubbornly resistant to proof. Some progress was achieved in 2004 when American mathematician Daniel Goldston and Turkish mathematician Cem Yildirim showed that, by making certain assumptions, including the truth of another unproven conjecture known as the Elliott–Halberstam conjecture, there were infinitely many prime pairs that differed by just 16. A flaw in their proof was corrected the following year with help from Hungarian mathematician János Pintz. Then, out of the blue, in 2013, a Chinese-born American maths professor at the University of New Hampshire, Yitang Zhang, made a startling announcement. He'd established, *without the need to make any other assumptions at all*, the first finite bound on the least gap between consecutive primes that is attained infinitely often. His proof showed, unconditionally, that there are infinitely many pairs of prime numbers that differ by 70 million or less. Obviously, 70 million is slightly more than 2! But the point is that Zhang had shown a way for other mathematicians to make further progress on the conjecture. In 2014, using Zhang's approach, the upper bound was slashed to 246, or, if a generalised form of the Elliott–Halberstam conjecture is assumed, as low as 6.

Refinements have also been made to the upper bound that Graham's number describes, as we'll see. But, for the moment, let's focus on trying to comprehend the sheer scale of this number and the extent to which it outshines every other star in the numerical firmament that we've come across so far.

It was in conversation with Martin Gardner, one of the greatest writers on popular mathematics, that Ron Graham gave a simplified explanation of the number named after him. The original Graham's number wasn't explicitly identified in the 1971 paper and its description there would have been too difficult for non-mathematicians to follow. Effectively, Graham invented a new number by a relatively easy-to-follow method that gave a feel for the enormous size of the original. This invented number was not only specific but was actually larger than the original so that it truly was an upper bound to the solution of Graham and Rothschild's problem! Gardner then made the new, easier-to-understand Graham's number the subject of his monthly column in the November 1977 issue of *Scientific American*. Three years later it won even more public fame with an entry in *The Guinness Book of Records* for being the largest number (at the time) ever used in a mathematical proof.

Graham's (new) number is nothing more than a power tower of 3s. In other words, it's 3 to the 3 to the 3 and so on. Written as a repeated exponential it looks like this:

where the dots stand for lots of 3s – in fact, a power tower of 3s so high that it's impossible to imagine. Even if the 3s were microscopic in size the tower would rise above the height of Mount Everest, and then on and on, far beyond the extent of the observable universe. We have to accept, from the outset, that we can never, ever truly grasp the size of Graham's number. All we can do is try to follow how it comes about.

The first step in reaching Graham's number is to think of the number 3 ↑↑↑↑ 3. Remember from the last chapter that one up-arrow, on its own, represents exponentiation, so that $3 \uparrow 3 = 3^3$. Two up-arrows represent the next rung up the hyperoperator ladder – tetration, or repeated exponentiation.

$$3 \uparrow\uparrow 3 = 3 \uparrow (3 \uparrow 3) = 3 \uparrow 27 = 7,625,597,484,987$$

Already, we've gone past 7 trillion. Next up is pentation:

$$3 \uparrow\uparrow\uparrow 3 = 3 \uparrow\uparrow (3 \uparrow\uparrow 3) = 3 \uparrow\uparrow 7,625,597,484,987$$

giving us a power tower that's 7.6 trillion threes high:

$$3^{3^{3^{\cdot^{\cdot^{3}}}}} \left.\right\} \text{7,625,597,484,987 high}$$

If we were to print this out, replacing the dots with the correct number of 3s, and assuming that each 3 was 3 millimetres high, it would stretch about 23 million kilometres into space. That's more than half the distance to Venus at its closest approach to Earth, so we might call this number the 'Venus tower'. Now add *another* up-arrow:

$$3 \uparrow\uparrow\uparrow\uparrow 3 = 3 \uparrow\uparrow\uparrow (3 \uparrow\uparrow\uparrow 3)$$

This takes us up to the level of hexation. Notice that inside the right-hand set of brackets is our Venus tower – a stack of threes that would reach over halfway to the nearest planet. So we can write:

$$3 \uparrow\uparrow\uparrow\uparrow 3 = 3 \uparrow\uparrow\uparrow \text{Venus tower}$$

What does this mean? Pentation (shown by three up-arrows) is repeated tetration, so that:

$$3 \uparrow\uparrow\uparrow \text{Venus tower} = 3 \uparrow\uparrow (3 \uparrow\uparrow (\dots \uparrow\uparrow 3))$$

Now we're dealing with not just a Venus power tower of threes but a Venus power tower of repeated tetrations! If we represent this vertically rather than horizontally, it looks like this:

$$3 \uparrow\uparrow\uparrow \text{Venus tower} = \left. \begin{array}{l} \underbrace{3^{3^{\cdot^{\cdot^{3}}}}} \\ \underbrace{3^{3^{\cdot^{\cdot^{3}}}}} \\ \vdots \\ 3 \end{array} \right\} \begin{array}{l} \text{(Venus tower} \\ -1\text{) towers} \end{array}$$

To be clear, this means taking the number 3, then taking a power tower of threes of height 3 to get roughly 7.6 trillion, then taking *this* result and making it the height of a power tower of threes to get the Venus tower, then taking *this* result and making it the height of a power tower of threes to get something vastly larger, and repeating the whole process a Venus tower number of times. There's no point in wearing out your neurons trying to grasp how big the result will be – you can't. No one can. You might just be able to picture in your mind's eye flying in a spaceship more than half the way to Venus and seeing, through the window, all those tiny threes marching across millions of kilometres of space and forming a power tower. But if you want to wrap your brain around the number (3 ↑↑↑ Venus tower), you have to picture each of those little threes being replaced by a power tower of threes.

So we've arrived at the first step along the way to Graham's number:

$$3 \uparrow\uparrow\uparrow\uparrow 3 = \underbrace{3 \uparrow\uparrow (3 \uparrow\uparrow (\ldots \uparrow\uparrow 3))}$$

Venus tower of a Venus
tower of tetrated 3s

Following the lead of Ron Graham and Martin Gardner, let's call this number g_1. Of course, it's shockingly big in terms of anything the physical cosmos has to offer. But here's the thing. In the next step along the road to Graham's number, this brain-bustingly large number, g_1, becomes merely the *number of up-arrows* used in defining a far, far larger number, g_2:

$$g_2 \longrightarrow \underbrace{3 \uparrow\uparrow \cdots\cdots \uparrow\uparrow 3}$$

$$g_1 \longrightarrow 3 \uparrow\uparrow\uparrow\uparrow 3$$

And, before you get comfortable with that idea, g_2 is the number of up-arrows in the definition of an even larger entity, g_3. On and on this insanity continues until, finally, we reach g_{64}, which is Graham's number!

$$g_{64} = \text{Graham's number} \longrightarrow \underbrace{3 \uparrow\uparrow \cdots\cdots\cdots\cdots \uparrow\uparrow 3}$$

$$g_{63} \longrightarrow \underbrace{3 \uparrow\uparrow \cdots\cdots\cdots \uparrow\uparrow 3}$$

$$\vdots$$

$$g_2 \longrightarrow \underbrace{3 \uparrow\uparrow \cdots\cdots \uparrow\uparrow 3}$$

$$g_1 \longrightarrow \underbrace{3 \uparrow\uparrow\uparrow\uparrow 3}$$

64 layers

Each g-number serves to specify how many up-arrows are needed to represent the next g-number in the series, all the way up to g_{64}. The growth rate of the g-numbers is astounding, incomprehensible. Graham's number itself is so large that, in comparison, Skewes' number and even Moser's number are utterly insignificant. The speed at which the g-numbers grow, from one to the next, and the ease with which, after a few dozen steps, we can arrive at something of the monstrous scale of g_{64} is testimony to the power of a process that we're going to see a lot more of in the chapters ahead – recursion.

Graham's number came about as an easier-to-digest version of the smallest known upper bound to the solution of the geometric Ramsey problem we talked about earlier. But in the years since it became famous it's been replaced in that role by successively smaller numbers. A 2014 paper by Mikhail Lavrov, Mitchell Lee, and John Mackey gave a new upper bound of 2 ↑↑↑ 6 – still a big number, though ludicrously small compared to g_{64}. Five years later, Eryk Lipka reduced the upper bound still further, to 2 ↑↑↑ 5. The true solution of the problem is thought to be very small indeed and probably close to the currently known lower bound, which, according to an online preprint submitted to Cornell's arXiv website in 2008, is 13.

We've seen that there's nowhere near enough room or matter in the observable universe to be able to write out in full some of the big named numbers that we've already come across – the googolplex, Skewes' number, Moser's number and so on. So obviously there's no chance that there'll ever be a complete digital representation of Graham's number, even if we knew all of its digits (which we never will) and even if every single Planck volume in the known universe were devoted to the task. But it's worse than that. If you think about the googolplex, it's impossible to write out in full but it's easy to write out in full *how many digits it contains* (just one followed by 100 zeros). In the case of Graham's number the observable universe isn't big enough to hold even the number of digits in the digital representation. Nor is it big enough to represent the number of digits representing the number of digits in the digital representation, and so on for a few dozen more such embeddings. We'll never be able to see Graham's number in full nor know what the vast bulk of its digits are but we can at least console ourselves by seeing how it ends: ...262464195387.

You may have noticed that Knuth's up-arrows have met their match in Graham's number. As formidable as this system of notation is compared with more familiar forms such as indices and power towers, it doesn't cope well with the kind of numerical giants we encounter once recursion takes hold. In the case of Graham's number we ended up using rows of dots to indicate where there were far too many up-arrows to write out in full. Fortunately, other ways of representing big numbers have been devised that can take over when up-arrows are overrun.

CHAPTER 6

Conway's Chains

ANYONE VENTURING INTO the universe of large numbers faces challenges like those of long-range space missions. As we seek to explore further and further from our home planet, we need to know what's out there and how to cross vast distances quickly, on a human timescale. For the foreseeable future, the planets and other objects in the Solar System are our main goals, but beyond them lie the many wonders of interstellar and then extragalactic space.

Already some of our spacecraft have left the Solar System and are heading towards the stars. *Voyagers 1* and *2*, launched in 1977, have passed through the outer boundary of the Sun's magnetic influence (the heliopause) and are now officially in interstellar space. Both are still sending back scientific data and, with any luck, will continue to do so for a few more years yet. After that they'll fall silent and we'll hear no more from them as they drift silently past some of the nearer stars before plunging deeper into the great voids of the Milky Way. *Pioneers 10* and *11*, launched earlier than the Voyagers but travelling more slowly, are also on exit trajectories from the Solar System. However, their power sources are already exhausted so that we've no way of communicating with them. Calculations by a team of researchers, published in 2019, suggest that none of these four spacecraft are likely to have any really close encounters with other star systems over the next million years or so. The nearest thing to a stellar 'flyby' may be when *Pioneer 10* comes within three quarters of a light-year of the orange dwarf star HIP 117795, in the constellation Cassiopeia, about 90,000 years from now.

It's hard to get excited about a remote stellar flyby by an inert spacecraft that won't happen for another 3,000 generations. To make interstellar space travel meaningful and interesting we need to develop methods of propulsion that

FIGURE 6.1: Artist's impression of *Pioneer 10* looking back on the Solar System while on its way to interstellar space.

will get us, or our robot probes, to other stars in a matter of decades, not tens of thousands of years. In the same way, when we're dealing with the space of large numbers we can't rely on traditional and hopelessly pedestrian means of travelling down the number line to reach targets such as Graham's number or whatever may lie still further away.

Graham's number is like an interesting but distant star. We know it's out there and that, given enough time, we could reach it one baby step at a time – in the mathematical case, simply by counting, 1, 2, 3, and so on, or by climbing up, one by one, the many levels of a fantastically tall power tower. But such conventional methods aren't practical for high-speed numerical travel any more than chemical rockets are up to the task of hopping between stars in the galaxy. Knuth's up-arrows (or their equivalent in H-operator or square-bracket notation) give us a form of propulsion for racing down the number line much faster than anything we learned about in school. Just a handful of up-arrows are needed to represent numbers so large that they'd fill the observable universe and more if written out in full. Yet, when confronted with Graham's number, even up-arrow propulsion turns out to be pitifully inadequate. We're forced to use dots to stand for rows of up-arrows so fantastically long that they couldn't be contained, however written, within the physical confines of space as we know it.

So, where do we go from here? What's the next great leap in numerical propulsion technology? It's worth recalling that up-arrows were Donald Knuth's way of showing hyperoperators, and hyperoperators, in turn, extend the hierarchy of operators with which we've been familiar since childhood: succession (better known as 'counting up by one'), addition, multiplication, and exponentiation. Apart from succession, which is a *unary* operation because it acts on just a single operand (the number that 1 is being added to), all the hyperoperators are binary or dyadic. For instance, $a + b$ and a^b are both binary operations because they act on two operands: a and b. The hyperoperator sequence is also recursive because each operator can be defined and represented in terms of the repeated action of the previous operator in the sequence.

To develop a propulsion system that will take us further and faster in space than ever before, we have to come up with an entirely new technology or make serious improvements to what's already available. What leap in mathematical engineering could hurl us at vastly greater speed towards far-flung corners of the numerical cosmos? In 1996, British mathematicians John Conway and Richard Guy, in their *Book of Numbers*, explained a new way of travelling down the number line by a method that, though based on up-arrows, was enormously more powerful. Like Donald Knuth, Conway and Guy, who both died in 2020, were eminent, professional mathematicians who also thoroughly enjoyed the recreational aspects of their subject and were quite happy to forge links between the academic and the more playful. Not surprisingly this brought them into close contact with American maestro of recreational maths writing, Martin Gardner, who then broadcast their discoveries to a worldwide audience.

John Horton Conway was born in Liverpool, in 1937, and showed an early aptitude for maths, reciting powers of two when he was just four years old. By the age of eleven he knew exactly what he wanted to do when he got older: study

FIGURE 6.2: John Horton Conway in 2005.

mathematics at Cambridge and go on to become a full-time academic mathematician. At school he was painfully intro-verted but at university managed to transform himself into such an extrovert for the rest of his life that a 2015 article on him in the *Guardian*, by Siobhan Roberts, was titled: 'John Horton Conway: the world's most charismatic mathema-tician'. He was known to be witty, boisterous, polymathic, and endlessly creative. Sir Michael Atiyah described him as 'the most magical mathematician in the world'.

Having earned his doctorate at Cambridge, in 1964, Conway stayed on as a member of the faculty there until, in 1986, he moved to Princeton, New Jersey, to become John von Neumann Professor of Applied and Computational

Mathematics. He held that professorship, later becoming emeritus, until his death in April 2020, aged eighty-two, of complications from Covid-19.

It's hard to draw a line between Conway's achievements in mainstream mathematical research and the many fascinating discoveries and inventions he made at the fringes of the subject. On the recreational front he was probably best known for his Game of Life, introduced to the world at large through one of Gardner's *Scientific American* columns in 1970. But Life is far more than just a pastime involving counters on a board marked with squares. It's one of the earliest successful examples of what's known as a cellular automaton – a way of modelling the evolution of a system that can be represented using a regular grid of cells. We'll have more to say about it in Chapter 8.

In the same year that the Game of Life became the thinking person's computer recreation of choice, Conway also provided the first description of surreal numbers. This new, vast number system, which, as we saw earlier, Donald Knuth explained in the guise of a fictional tale, stemmed, oddly enough, from Conway's analysis of endgame situations in the ancient board game of Go. Conway made important contributions to many areas of maths, including theories of groups, knots, and games, geometry, algebra, and topology. He co-authored several books with fellow number theorist and recreational mathematician Richard Guy, who remained active in his academic fields, as well as environmentalism and mountain hiking, up until his death, just a month or so before Conway's, at the age of 103. In one of the books they wrote together, Conway unveiled a new, ingenious way of representing large numbers called chained-arrow notation.

In the last chapter we used up-arrows to get to Graham's number, g_{64}. But long before we reached our goal it was obvious that Knuth's notation wasn't really up to the task. It was as inadequate for the job as a crewed spaceship to Alpha Centauri fitted with the same kind of rockets that sent humans to the Moon. At first sight, Conway's system for representing large numbers, which uses right-pointing arrows, doesn't seem much of an improvement, if any, over up-arrows.

Like one of Knuth's chains, a Conway chain, such as $3 \rightarrow 2 \rightarrow 3 \rightarrow 4$, consists of positive integers separated by arrows. Also like up-arrows, the sole purpose of chained arrows is to provide a compact means of representing extraordinarily large numbers. A chain of length 1 (so, not really a chain at all!) is just a positive whole number, and a chain of two numbers is equivalent to one up-arrow or exponentiation:

$$a \rightarrow b = a \uparrow b = a^b$$

A chain of three numbers, $a \rightarrow b \rightarrow c$, is equivalent to a hyperoperation or multiple up-arrows:

$a \rightarrow b \rightarrow c = a \uparrow^c b$ (in other words, a raised to a power tower of bs that is c high).

You may be thinking at this point that Conway's and Knuth's systems are, in fact, identical, except that the arrows point in different directions! But the resemblance is superficial. The immense difference between the two becomes apparent when we start to look at chains that are four or more in length, for instance:

$3 \rightarrow 4 \rightarrow 2 \rightarrow 6$, or

$6 \rightarrow 5 \rightarrow 10 \rightarrow 5 \rightarrow 5$

You can figure out what any number written in chained-arrow form is by using just a few simple rules, a couple of which we've already met. There are various ways of writing these rules, all of which amount to the same thing, and we'll just choose a set that's easy to apply. Rule 1 is the trivial statement that if a chain a is only one element long, then it's just the number a. Rule 2 is that a chain of length two is equivalent to a single up-arrow or exponentiation: $a \rightarrow b = a \uparrow b = a^b$. Rule 3 says that if a chain ends in 1, you can just drop the 1 because it's not doing anything. For example, $4 \rightarrow 5 \rightarrow 8 \rightarrow 4 \rightarrow 1$ is exactly the same as $4 \rightarrow 5 \rightarrow 8 \rightarrow 4$. In the same way, Rule 4 says that if the *second* last number is 1, you can chop off the chain at that point, so that, for instance $3 \rightarrow 6 \rightarrow 1 \rightarrow 4$ immediately reduces to $3 \rightarrow 6$. Rule 5 is what gives Conway's chained arrows their phenomenal power and the one that takes a bit to get your head around. We can write it like this:

$$X \rightarrow a \rightarrow b = X \rightarrow (X \rightarrow (a-1) \rightarrow b) \rightarrow (b-1)$$

where X is the chain up as far as the last two numbers, and a and b are the last two numbers. In plain English, this means we decrease the last number by 1 and feed the original chain with the second last number decreased by 1 into the second last number. In other words, we replace the b in the original chain with $(b-1)$ and substitute the a in the original chain with another copy of the original chain, except that the a in this copy is replaced by $(a-1)$. Then we just repeat the process over and over again, always evaluating what's inside the innermost set of parentheses before moving on to the rest of the chain. Here's a summary of the five rules:

1   A chain a of length one is just the number a.
2   $a \rightarrow b$ is the same as exponentiation, a^b, or a single up-arrow, $a \uparrow b$.
3   $X \rightarrow 1 = X$, where X is a sub-chain, or part of a chain. In this case, X is all of the chain except for the last element.
4   $X \rightarrow 1 \rightarrow b = X$.
5   $X \rightarrow a \rightarrow b = X \rightarrow (X \rightarrow (a-1) \rightarrow b) \rightarrow (b-1)$

Let's now look at a few simple examples of how these rules work together, starting with the three-number chain $3 \rightarrow 3 \rightarrow 2$. Applying Rule 5, with $X = 3$, $a = 3$, and $b = 2$, we get:

$$3 \rightarrow 3 \rightarrow 2 = 3 \rightarrow (3 \rightarrow 2 \rightarrow 2) \rightarrow 1$$

Rule 3 lets us drop the 1 at the end, so this becomes:

$$3 \rightarrow (3 \rightarrow 2 \rightarrow 2)$$

Now, let's focus on what's inside the parentheses and apply Rule 5 to it:

$$3 \rightarrow (3 \rightarrow (3 \rightarrow 1 \rightarrow 2) \rightarrow 1)$$

Drop the 1 at the end and the chain becomes:

$$3 \rightarrow (3 \rightarrow (3 \rightarrow 1 \rightarrow 2))$$

Rule 4 allows the $3 \to 1 \to 2$ inside the parentheses to be truncated to 3 on its own, so we're left with $3 \to (3 \to 3)$. The $3 \to 3$ inside the remaining parentheses is, by Rule 2, just 3 cubed or 27, so the whole chain boils down to 3^{27}, or 7,625,597,484,987. This agrees with what we said earlier about the interpretation of a three-number chain in Conway's notation: namely, that it's the same as a hyperoperation with the last number in the chain giving the number of up-arrows. In other words:

$$3 \to 3 \to 2 = 3 \uparrow\uparrow 3 = 3 \wedge 3 \wedge 3 = 3^{27}$$

The size of numbers represented in chained-arrow form depends greatly on both the length of the chain and the size of the individual elements. The four-number chain $2 \to 3 \to 2 \to 2$, for instance, reduces, according to Rule 5, after the first step to:

$$2 \to 3 \to (2 \to 3) \to 1$$

which then becomes:

$$2 \to 3 \to 8$$

This is equivalent, as we know, to $2 \uparrow\uparrow\uparrow\uparrow\uparrow\uparrow\uparrow\uparrow 3$, with eight up-arrows. It's a colossal number by any normal standards but not in the same league as, say, Graham's number. Consider, though, the chain $5 \to 4 \to 3 \to 3$. Rule 5 tells us this is the same as:

$$5 \to 4 \to (5 \to 4 \to 2 \to 3) \to 2$$

The next step is to begin to evaluate what's inside the parentheses, which again involves applying Rule 5:

$$5 \to 4 \to (5 \to 4 \to (5 \to 4 \to 1 \to 3) \to 2) \to 2$$

Now we have nested parentheses, and we have to focus on figuring out the inner one before doing anything else. By Rule 4, the chain in the inner parenthesis can be chopped off at the 1, leaving us with

$$5 \to 4 \to (5 \to 4 \to (5 \to 4) \to 2) \to 2$$

The $5 \to 4$ in the inner parenthesis is $5^4 = 625$, so now we have:

$$5 \to 4 \to (5 \to 4 \to 625 \to 2) \to 2$$

and already we can sense that there's about to be a major eruption because of the sudden appearance of that 625 in the middle of the chain. Applying Rule 5 to the contents of the remaining parenthesis:

$$5 \to 4 \to (5 \to 4 \to (5 \to 4 \to 624 \to 2) \to 1) \to 2$$

and suddenly we're riding the interstellar chained-arrow-powered hyperdrive to some number that before we had no means to reach. The next step in the expansion gives:

$$5 \to 4 \to (5 \to 4 \to (5 \to 4 \to 624 \to 2)) \to 2$$

The inner section then reduces to:

$$5 \to 4 \to (5 \to 4 \to (5 \to 4 \to (5 \to 4 \to 623 \to 2) \to 1)) \to 2$$

After hundreds more iterations, the third element in the innermost set of parentheses (currently 623) will finally have shrunk down to 1 and we'll be able to start on the next level, which will again contain the chain $5 \to 4 \to 625$. By this time, however, the overall chain will be nearly two thousand elements long, and about to grow much longer still. Now we can start working outwards, but at each step the value we get becomes the number of up-arrows in the next, just as in the expansion of Graham's number. Finally, once all the inner parentheses are resolved, the result we end up with is then the third element of another chain ending with 2. This means we have to repeat the same process of expansion, but now the number of steps we take, each step becoming the number of up-arrows in the next, is equal to this gigantic number we already formed using this process!

The chained-arrow $5 \to 4 \to 3 \to 3$ is, in fact, much larger than Graham's number, which lies between $3 \to 3 \to 64 \to 2$ and $3 \to 3 \to 65 \to 2$. You might think, at first glance, that $3 \to 3 \to 64 \to 2$ is bigger than $5 \to 4 \to 3 \to 3$ because 64 is by far the largest single element in either chain. But having a 3 at the end of a chain rather than a 2 is the vastly more dominant factor in determining how many steps it takes to write the chain out in full.

$5 \to 4 \to 3 \to 3$ makes Graham's number look insignificant. Yet we can easily name a bigger chain. How about $35 \to 269 \to 81 \to 95 \to 54 \to 428$? Written in this way it's as compact, in terms of the space it takes up on the page, as $35 + 269 + 81 + 95 + 54 + 428$, which we can easily calculate to be a mere 962. There's nothing about the appearance of a number in chained-arrow form that gives any inkling whatsoever of the size of the monster lying curled up within. But as soon as we start to unpack it, realisation dawns about what we're dealing with.

Having glimpsed the task of unravelling a chain even as simple-looking as $5 \rightarrow 4 \rightarrow 3 \rightarrow 3$, it might seem that evaluating a longer chain involving much larger elements would be impossible. And, indeed, it *is* impossible practically speaking. The whole point of chained-arrow notation is that it provides a means of representing numbers so huge that they *can't* be written in ordinary form! The universe isn't big enough to hold them. Furthermore, it's very easy to represent a number in chained-arrow form that's so large it couldn't be written in full even using up-arrows because it would require more up-arrows than there are Planck volumes in the observable universe. In fact, it's also easy and quick to come up with a Conway chain describing a number so big that even the *number of up-arrows* needed to represent it couldn't be written down, not if every Planck volume were used to hold a digit of it.

Yet, each and every Conway chain does represent a specific finite number, however large, and, if time, space, matter, and energy weren't limiting factors, the chain could eventually be evaluated and reduced to a single number. This guaranteed reducibility comes from the clever way in which the rules governing chained-arrow notation were devised. When the last number or the next-to-last number in a chain reduces to one, as it surely will, the chain can be truncated at that point, and the process continued until all the elements have been whittled down to the last one standing – the final result.

The ability of a very simple set of rules to give rise to such a powerful system for naming gigantic numbers parallels the similar simplicity of the Game of Life – another of Conway's inventions. In Life the rules are so elementary that a child can follow them, yet it can lead to patterns of astonishing diversity and complexity. In chained-arrow notation all the potential for concealing super-scale numbers behind a veneer of simplicity is contained in what we've called Rule 5.

As in the case of the hyperoperator sequence, the key force at work in Conway's system is recursion – the repeated application of a rule or formula to its own results. In a broader sense, we can see recursion at work in the everyday world when, for instance, we stand between two parallel mirrors: our image in one mirror appears in the image of the other, whose image in turn becomes part of the image in the first, and so on. A well-known toy – the Russian doll – is recursive in that a small doll lies inside a larger one, which in turn is concealed within a still larger one.

We use recursion in maths right from the start, as young children, when we learn counting. In formal terms, counting is just the repeated application of the successor function, the action of which is to add one to the last result. Repeated counting is addition, repeated addition is multiplication, repeated multiplication is exponentiation, and so on, so that each step up the hyperoperator sequence marks a new level of recursive processing. Chained arrows harness even more recursive power to express numbers that, realistically, are beyond the reach of up-arrows or any of the other schemes for representing hyperoperators. But Conway's arrows themselves are little more than a playful game – a trick devised by a skilful mathematician to show us what amazing things recursion can do in the realm of large numbers. Now it's time to move beyond the 'popular' big-number schemes, such as up-arrows and chained arrows. A journey into some deep and serious mathematics awaits us on the way to the biggest number in the world – a journey that began almost a century ago.

CHAPTER 7

Ackermann and the Power of Recursion

THE FIRST QUARTER of the twentieth century was a time of turmoil and revolution in almost every aspect of society, including science and mathematics. Just as physics had been rocked by quantum theory and relativity, mathematicians, too, were questioning the very foundations of their subject. Was all of mathematics provable starting from a basic set of assumptions or axioms? Among those who believed so was the preeminent theorist of his day, David Hilbert. In 1900, at the International Congress of Mathematicians in Paris, Hilbert announced what he considered to be the 23 greatest unsolved problems in maths. The second of these bore directly on his most fervent belief: that through logic everything that was true in mathematics could be rigorously proved to be true from the initial assumptions.

In 1920, Hilbert proposed a research project, which became known as Hilbert's Program, to reformulate the whole of maths on a solid, logical foundation. At the time, one of his postgrad students was twenty-four-year-old

Wilhelm Ackermann. Born in Schönebeck, Westphalia, Ackermann enrolled as a teenager at the University of Göttingen, then the world's foremost institution for mathematics and where Hilbert was a professor. The First World War interrupted his studies, which focused on maths, physics, and philosophy, but eventually, in 1925, Ackermann earned his PhD and then spent some time at Cambridge on a scholarship. Shortly after, he got married, which didn't go down well with Hilbert, who thought that young researchers should stay single and devote themselves exclusively to their chosen field. The final straw came when he found that the Ackermanns were expecting a child.

'Oh that is wonderful!' Hilbert said sarcastically. 'That is wonderful news for me. Because if this man is so crazy that he gets married then even has a child, it completely relieves me from having to do anything for such a crazy man!'

The fact that Hilbert himself had married and become a father when only a few years older than Ackermann didn't seem to come into his reckoning. Though doubtless a first-rate genius, Hilbert seems also to have been something of an intellectual snob. On another occasion, after hearing that one of his students had dropped out to study poetry, he remarked: 'Good, he did not have enough imagination to become a mathematician.'

Hilbert's ostracising of Ackermann, despite the young man's brilliance as a logician, got in the way of him obtaining a university position and he ended up teaching in high schools for the next thirty-four years. Despite this handicap, Ackermann remained a major force in the field of mathematical logic. In 1953 he became a member of the Göttingen Academy of Sciences and was made an honorary professor in the School of Mathematics and Sciences at Münster University. He delivered a lecture there just three days before his death on Christmas Eve, 1962.

Ackermann had been working as Hilbert's personal secretary before the great man fell out with him for daring to start a family. In fact, the two wrote an important textbook together called *Principles of Mathematical Logic*, which was published in Germany in 1928. Based on lectures that Hilbert delivered at Göttingen between 1917 and 1922, it's considered to be the first exposition of what's known as first-order logic (FOL). In this type of logic, every statement can be expressed, as in the case of an ordinary sentence, in the form of a subject and a predicate. The subject is whatever is being discussed or described and the predicate is what we're saying about the subject. The only difference is that in mathematical logic we use special symbols and abbreviations rather than words. For example, if we let 'A' stand for apples and 'F' for fruit, then in first-order logic we could write the statement: $\forall x : Ax \rightarrow Fx$. In plain English, this means: 'For all x, if x is an apple, then x is a fruit'.

Hilbert and Ackermann's book offered the first thorough grounding in FOL but it went far beyond simply explaining how this most fundamental of symbolic logic systems worked. It also examined whether first-order logic was complete – in other words, whether all true statements in FOL could be derived from the axioms that underpin it. In so doing it introduced the world to the *Entscheidungsproblem*, or decision problem, which we met in Chapter 5. Today FOL is widely used in proof theory (the study of how to prove things formally in maths) and other areas at the foundations of mathematics.

In the late 1920s, around the time *Principles of Mathematical Logic* was being written, Ackermann and Romanian mathematician Gabriel Sudan, who was also a student of Hilbert's, were working on the fundamentals of another subject that underpins maths – computability theory. Today, part of computability theory overlaps with computer science, which is concerned with the design of software systems and how data can be handled by machines many times faster than by humans. But the first aspects of computability theory to be developed had nothing to do with practical data processing. Instead, they grew out of mathematical logic, more than a decade before the first true computers appeared on the scene. Computability theory, in the broadest sense, has to do with what it's possible to compute or evaluate and what, even in principle, is non-computable. It's also known as recursion theory, which gives a clue to its central relevance to the subject of big numbers.

The basic objects with which computability theory deals are known as computable functions. We're going to be talking a lot about functions, of various kinds, in what lies ahead. So, before moving on, let's pause and make sure we're clear on exactly what a function is in mathematics. One way to think about a function is in terms of a little machine or black box that turns one set of numbers into another, so that there's always a one-to-one correspondence between the set of numbers going in and the numbers coming out. Alternatively, you can think of a function as a rule for bringing about such a transformation, or a relationship that ties every member of one set to every member of another. If x is our variable, or set of input numbers, then we can write 'a function of x' as $f(x)$. It might be, for instance, that $f(x) = 2x + 1$, which means that we double each input value of x and add 1 to it to get the output – in other words, the corresponding value of $f(x)$. Graphed as x against $f(x)$, the result will be a straight line. On the other hand, the function might be $f(x) = x^2$, so that each value of x must be multiplied by itself to produce the associated value of $f(x)$. In this case, the graph will be a roughly U-shaped curve called a parabola.

A computable function is one that can be expressed in the form of an algorithm, or a sequence of precise instructions. This definition includes all of the functions that most of us will ever come across in mathematics at school or beyond. What's more, all the functions we usually meet are also of a type said to be 'primitive recursive'. Addition, multiplication, exponentiation, and the factorial function ($n! = n \times (n - 1) \times \ldots \times 3 \times 2 \times 1$), to name but a few, are among the functions that fall into this category. One way to think of a primitive recursive function is in terms of computer programs – although the theory of them was developed long before computers were invented.

A common feature of many computer programs is the loop. This is a piece of code that tells the computer to keep going round and round through a series of instructions, either a specified number of times or until a certain condition is met. A loop where the number of repeats is fixed at the outset is called a count-controlled loop or 'for loop'. It might look like this, for instance:

```
FOR I = 1 TO N
    xxx
NEXT I
```

where xxx is the body of the loop.

The other two main types of loop are the condition-controlled varieties, known as 'while loops' and 'do-while loops'. In these, passages through the loop continue only until some circumstance arises. The condition may be tested at the start of each loop or at the end.

If we think in terms of loops, a primitive recursive function is one that can be computed running a program in which the only loops are of the count-controlled or 'for' type. In other words, the upper bound of the number of times the loop has to be gone through must be specified in advance. What Ackermann and Sudan were trying to come up with was a computable function that *wasn't* primitive recursive. Both succeeded but in slightly different ways.

Gabriel Sudan discovered and published his function first, in 1927. However, it's the Ackermann function, formulated the following year, which is now much better known and which we'll look at in more depth here. Ackermann proved the non-primitive nature of his creation in a paper called 'On Hilbert's Construction of the Real Numbers', published in 1928. In its original form, the Ackermann function was denoted by the Greek letter phi (φ) and depended on three input values or 'arguments', so that it could be written, for instance, as $\varphi(m, n, p)$. Ackermann defined it using five rules. In later years, other mathematicians, including the Hungarian Rózsa Péter and the American Raphael Robinson, revised and simplified the original Ackermann function to make it easier to work with. When the Ackermann function is referred to today it's usually in the form of one of these streamlined variants. The specific one we'll look at here is the Ackermann–Péter function, $A(m, n)$.

The Ackermann–Péter function depends on just two variables, m and n, both of which can have only non-negative, whole number values. It's defined by just three rules:

$A(0, n) = n + 1$
$A(m, 0) = A(m - 1, 1)$ for m greater than zero
$A(m, n) = A(m - 1, A(m, n - 1))$ for m and n both greater than zero

Remember, this is a function constructed for the sole purpose of providing an example of a computable function that's not primitive recursive. With that in mind, let's put in a few different values for m and n and see what happens.

The first rule is straightforward enough. If we put $n = 2$ and apply the first rule, we get:

$A(0, 2) = 2 + 1 = 3$

In other words, the value of the Ackermann–Péter function when $m = 0$ and $n = 2$, $A(0, 2)$, is 3. The second rule is more interesting because the function A appears on both sides of the equality. If we put $m = 2$ in it, we get:

$A(2, 0) = A(1, 1)$

Now what? Neither of the first two rules covers this situation – where both the arguments of the function are nonzero. We have to apply Rule 3:

$A(1, 1) = A(0, A(1, 0))$
$= A(0, A(0, 1))$ (applying Rule 2 to figure out $A(1, 0)$)

$= A(0, 2)$ (applying Rule 1 to evaluate $A(0, 1)$)
$= 3$ (again by Rule 1)

At this stage, it seems that turning to Ackermann's function in our search for the biggest number in the world is proving to be a bit of a let-down! We've gone from Conway's mighty chained arrows, and numbers so big that no superlative is adequate to describe them, to 3!

Both $A(0, 2)$ and $A(2, 0)$ boil down to the number 3. How about $A(2, 2)$? In this case, the evaluation process takes a little longer – 27 steps in all. Feel free then to skip over the intermediate workings shown here and go straight to the punch line below.

$A(2, 2)$
$= A(1, A(2, 1))$
$= A(1, A(1, A(2, 0)))$
$= A(1, A(1, A(1, 1)))$
$= A(1, A(1, A(0, A(1, 0))))$
$= A(1, A(1, A(0, A(0, 1))))$
$= A(1, A(1, A(0, 2)))$
$= A(1, A(1, 3))$
$= A(1, A(0, A(1, 2)))$
$= A(1, A(0, A(0, A(1, 1))))$
$= A(1, A(0, A(0, A(0, A(1,0)))))$
$= A(1, A(0, A(0, A(0, A(0, 1)))))$
$= A(1, A(0, A(0, A(0, 2))))$
$= A(1, A(0, A(0, 3)))$
$= A(1, A(0, 4))$
$= A(1, 5)$
$= A(0, A(1, 4))$

$$= A(0, A(0, A(1, 3)))$$
$$= A(0, A(0, A(0, A(1, 2))))$$
$$= A(0, A(0, A(0, A(0, A(1, 1)))))$$
$$= A(0, A(0, A(0, A(0, A(0, A(1, 0))))))$$
$$= A(0, A(0, A(0, A(0, A(0, A(0, 1))))))$$
$$= A(0, A(0, A(0, A(0, A(0, 2)))))$$
$$= A(0, A(0, A(0, A(0, 3))))$$
$$= A(0, A(0, A(0, 4)))$$
$$= A(0, A(0, 5))$$
$$= A(0, 6) = 7$$

And there you have it: $A(2, 2) = 7$. It seems like we've expended an awful lot of effort for very little gain in terms of reaching big numbers. The first three values of the Ackermann function we've bothered to calculate are 3, 3, and 7. Moving on, $A(3, 2)$ turns out to be 29. How about $A(4, 2)$? If you thought the workings out for $A(2, 2)$ were long and laborious, we'll spare you the details of $A(4, 2)$! Suffice it to say that if you *did* have the patience and time to go through all the stages of figuring out $A(4, 2)$, you'd end up with a number that was 19,429 digits in length! The results of evaluating the Ackermann function for specific values of m and n grow very large, very quickly as m and n are increased, especially when m is 4 or greater.

There's a reason for this sudden explosive growth: the Ackermann–Péter function is none other than Knuth's up-arrows in disguise. Or, to be chronologically accurate, up-arrows are just a fun variation on the theme of the Ackermann function, which predates Knuth's scheme by almost half a century. Here's how the two are related:

$$A(1, n) = (n + 3) + 2 - 3 = n + 2 \qquad \text{(addition)}$$
$$A(2, n) = (n + 3) \times 2 - 3 = 2n + 3 \qquad \text{(multiplication)}$$
$$A(3, n) = 2^{n+3} - 3 \qquad \text{(exponentiation)}$$
$$A(4, n) = 2 \uparrow\uparrow (n + 3) - 3 \qquad \text{(tetration)}$$
$$A(5, n) = 2 \uparrow\uparrow\uparrow (n + 3) - 3 \qquad \text{(pentation)}$$

and so on.

The Ackermann–Péter function, it turns out, is effectively the same as up-arrows with a base of 2 (the number to the left of the up-arrows) and an offset of 3 (the −3 at the end). The table here shows how the function grows for values of m for 0 to 6 and n for 0 to 4. The m value is the dominant factor because it controls the number of up-arrows.

m \ n	0	1	2	3	4
0	1	2	3	4	5
1	2	3	4	5	6
2	3	5	7	9	11
3	5	13	29	61	125
4	13 $= 2^{2^2} - 3$ $= 2 \uparrow\uparrow 3 - 3$	65533 $= 2^{2^{2^2}} - 3$ $= 2 \uparrow\uparrow 4 - 3$	$= 2^{65536} - 3$ $= 2^{2^{2^{2^2}}} - 3$ $= 2 \uparrow\uparrow 5 - 3$	$= 2^{2^{65536}} - 3$ $= 2^{2^{2^{2^{2^2}}}} - 3$ $= 2 \uparrow\uparrow 6 - 3$	$= 2^{2^{2^{65536}}} - 3$ $= 2^{2^{2^{2^{2^{2^2}}}}} - 3$ $= 2 \uparrow\uparrow 7 - 3$
5	65533 $= 2 \uparrow\uparrow (2 \uparrow\uparrow 2) - 3$ $= 2 \uparrow\uparrow\uparrow 3 - 3$	$= 2 \uparrow\uparrow\uparrow 4 - 3$	$= 2 \uparrow\uparrow\uparrow 5 - 3$	$= 2 \uparrow\uparrow\uparrow 6 - 3$	$= 2 \uparrow\uparrow\uparrow 7 - 3$
6	$= 2 \uparrow\uparrow\uparrow\uparrow 3 - 3$	$= 2 \uparrow\uparrow\uparrow\uparrow 4 - 3$	$= 2 \uparrow\uparrow\uparrow\uparrow 5 - 3$	$= 2 \uparrow\uparrow\uparrow\uparrow 6 - 3$	$= 2 \uparrow\uparrow\uparrow\uparrow 7 - 3$

The original Ackermann function – the 1928 version – along with the Sudan function are the great-grandparents of all the ways we've looked at so far in this book for generating numbers too large to be represented by conventional mathematical means. At the same time, they're not as well known as the googol or googolplex, up-arrows, or chained arrows, all of which were introduced in a part-playful fashion by mathematicians to make big-number concepts accessible to a wider audience.

The Ackermann and Sudan functions were devised to answer serious questions in research mathematics. Their purpose, as we've seen, wasn't to provide a new way of representing big numbers. In fact, the original authors never mentioned, in their papers, how spectacularly fast the functions grew or anything at all about their capacity for generating gigantic numbers. That wasn't the object of what they were doing. Ackermann and Sudan's mission was pure and simple: to construct computable functions that weren't primitive recursive and to prove this fact about them to other mathematicians.

The various schemes we've explored in earlier chapters for representing immense numbers, including Steinhaus–Moser notation, Knuth's up-arrows, and Conway's chained arrows, all riff on the same underlying process that's at the heart of the venerable functions devised by Ackermann and Sudan – recursion. It's something we've been familiar with since we were very young, even if we were never introduced to the term itself. Children use recursion when they learn to count: 1, 2, 3, … We use it in addition, multiplication, and exponentiation. Recursion happens when something is defined or referenced in terms of itself. So, repeatedly adding

1 to whatever number came before, or repeatedly adding a number to itself, are recursive actions.

But there are different degrees or strengths of recursion. In terms of reaching big numbers quickly, multiplication is more powerful than addition (because it amounts to repeated addition), and exponentiation is more powerful than multiplication (because it's repeated multiplication). Tetration out-muscles exponentiation, and so on as we ascend the never-ending stairway of hyperoperators.

Although tetration, for instance, can produce big numbers faster than can exponentiation, and pentation is more powerful still, each one of the hyperoperators on its own falls within the same broad category of recursive functions: namely, the category we've identified as primitive recursive. In computer terms, each individual hyperoperator needs only a finite number of for loops to represent it – one for exponentiation, two for tetration, and so on.

We said earlier that Knuth's up-arrows are just another way of notating the hyperoperator sequence. For instance, $2 \uparrow 3$ is the same as 2^3 ($= 8$), two up-arrows is equivalent to tetration, etc. If the hyperoperators are primitive recursive, then so also are functions expressed in up-arrow form. Yet, we've seen that the Ackermann function, which is non-primitive recursive, can be written in terms of up-arrows! Haven't we just contradicted ourselves?

It's true that each of the hyperoperators *individually*, whether addition or something further up the ladder, such as hexation, is primitive recursive. But a function that can represent *all of the hyperoperators at once* – and there are infinitely many such operators – is an entirely different thing. This is effectively what the Ackermann function does. Its definition is sufficiently broad and powerful to encompass the entirety of the hyperoperator sequence. In terms of for loops, one for loop is needed to express $A(3, n)$, two in the case of $A(4, n)$, three in the case of $A(5, n)$, and so on. However, there's no way to express the full Ackermann–Péter function, which embraces every possible hyperoperator in the infinite sequence of hyperoperators, with finitely many for loops.

To harness more propulsive power from recursion than is available in any individual hyperoperator, the key is to feed a recursive function back into itself. Then it can effectively magnify its own strength, over and over again. In the case of the Ackermann–Péter function, this happens because of the third rule used in evaluating the function for specific values of m and n: $A(m, n) = A(m - 1, A(m, n - 1))$. On the right-hand side, the first argument – the m value – is decreased by one, but the new value for n is the Ackermann function fed back in with the same value of m and the n decreased by one. It's this repeated inputting of the Ackermann function from the previous cycle that's responsible for the function's explosive growth. At the same time, for any starting values of m and n, the Ackermann function has a built-in self-limiting feature. It's bound to produce a specific output in the end – a definite final answer – even if that number is phenomenally large. This is because each iteration involves either a smaller m, or an equal m and smaller n, so that the calculation is guaranteed to come to a conclusion eventually.

We've said that the Ackermann function and its slightly older cousin, the Sudan function, are the ancestors of all the various clever schemes later mathematicians have devised to explain how to reach numbers much bigger than those we'd otherwise encounter. Thanks to the likes of Steinhaus, Moser, Graham, Knuth, and Conway, and mathematical writers, such as Martin Gardner, who've provided accessible explanations for the lay reader, we've glimpsed ranges of numerical peaks that tower above the mere foothills of what we previously thought of as large numbers. Functions like those of Ackermann and Sudan are rarely mentioned at a popular level, whereas Knuth's and Conway's arrows, for instance, have had much more exposure in books and articles on recreational maths. What connects all these entities, however, is that no function which is primitive recursive can grow as fast as them.

Another point to notice about the various forms of non-primitive recursive functions that we've met, from Ackermann's original function to Conway's chained arrows, is that they're not binary. All the familiar operations of arithmetic, such as addition, multiplication, and exponentiation, act on just two values, or operands, at a time. If we call these operands a and b, then addition is $a + b$, multiplication is $a \times b$, and exponentiation is a^b. Of course, we can add, for example, more than two numbers in a row, but each individual addition is binary. This isn't the case with, say, the Ackermann function.

The Ackermann–Péter function, on the other hand, is just $A(m, n)$. It has only two variables, m and n, which makes it a binary function. So, how can it be essentially equivalent to Ackermann's original function, or to Knuth's up-arrows, when neither of those is binary? The answer is that the third variable in those functions is pinned down to the specific value of 2 in the Ackermann–Péter function.

This 2 appears as the base when the function is written in the form of Knuth's up-arrows:

$$A(m, n) = 2 \uparrow^{m-2} (n + 3) - 3$$

In terms of Conway's chained-arrow notation, the Ackermann–Péter function is equivalent to the three-element chain:

$$A(m, n) = (2 \to (n + 3) \to (m - 2)) - 3$$

in which, again, the 2 appears as the first element. Like the original Ackermann function, Conway's chained arrows are not binary in nature. In particular, when three or more numbers are joined by arrows the arrows don't act separately. Instead the whole chain has to be considered as a unit.

The Ackermann and other related non-primitive recursive functions are the jumping-off point for more developments in the use of specially crafted functions for reaching numbers at the edge of the observable universe of maths. But to understand these developments we need to delve further into computability theory and explore the limits of what it's possible, even in theory, to compute.

CHAPTER 8

Figure This – If You Can

FORGET THE LATEST high-end laptops with their terabytes of SSD and multi-gigahertz processor chips. Forget even the most powerful supercomputer in the world today – Fugaku, in Kobe, Japan, which can motor along at 415 petaflops – 415 thousand trillion floating-point operations a second. All you need to be able to compute anything that it's possible to compute, now or forever in the future, is a strip of paper, a read/write head – and a fair slice of patience. All you need, in fact, is a Turing machine.

Alan Turing was born in Maida Vale, London, in 1912, and even as a young child showed signs of the genius that would propel him to become a leading light in the new field of computer science. As an undergraduate at Cambridge, Turing took a course in logic during which he learned about the *Entscheidungsproblem*. He decided to focus on it as part of his graduate research having become convinced that Hilbert was wrong.

Remember, the *Entscheidungsproblem*, or 'decision problem', asks whether it's always possible to find a step-by-step procedure – an algorithm – to decide, in a finite amount of time, if a given mathematical statement is true or not. Hilbert strongly suspected that it was but he had no proof.

Obviously, there are a lot of possible mathematical statements you could make – in fact, infinitely many. So there's no chance of writing an algorithm to check every one, individually, to see if it ends in a true or false answer. What was needed, Turing realised, was some general way of implementing algorithms. Then he could perhaps put the decision problem to the test. He came up with the idea of a device that could carry out any logical set of instructions that was given to it. Although he called this device an *a*-machine (*a* for 'automatic'), others quickly began to refer to it after its inventor. Turing never intended that his machine should actually be built: it was meant as a purely abstract thing – no more than a mathematical model of a computing machine made from the simplest of components.

A Turing machine consists of just a read/write head and an indefinitely long tape divided into squares, onto each of which can be written a 1 or a 0, or which can be left blank. The head scans one square at a time and carries out an action based on the head's internal state, the contents of the square, and the current instruction in its logbook or program. The current instruction might be, for instance: 'If you're in state 12 and the square you're looking at contains a 1 then change it to a 0, move the tape one square to the left, and switch into state 23.'

On the machine's tape at the start is its input, in the form of a series of ones and zeros. The read/write head is positioned over the first square of the input, say the leftmost, and follows the first instruction that it's been given.

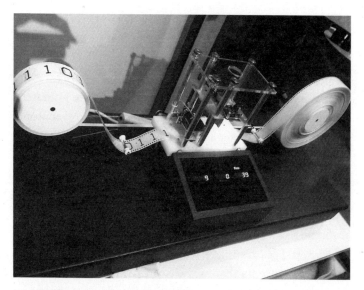

FIGURE 8.1: A working model of a Turing machine on display at the Harvard Collection of Historical Scientific Instruments.

Gradually, it works its way through the instruction list, or program, transforming the initial string of ones and zeros on the tape into a different string, until eventually it comes to a halt. When the machine reaches this final state, what's left on the tape is the output.

One of the simplest (and most useless!) tasks you could give a Turing machine would be to add one more to a row of 1s. The input would be the existing string of 1s followed by a blank square. The first instruction would tell the read/ write head to start at the first non-blank square and read what was on the square there. If it were a 1, the instruction would be to leave it unchanged and move one square to the right while remaining in the same state; if it were a blank, the instruction would be to write a 1 in that square and stop. If the head had advanced to the next square, the instruction would be repeated, and then again and again, until the head finally reached a blank and replaced it with a 1. Having added a 1 to the string, the head might be told to stop where it was or return to the start, possibly to repeat the whole process again and add one more to the total. Alternatively, a different state could be introduced when the read/write head is positioned at the final 1 and a new program of actions continued from there.

Some tasks given to a Turing machine might cause it to go on forever. For example, if you instructed a Turing machine always to move its read/write head to the right after each step, no matter what is on the current square, it will never stop, and it's easy to see in advance that this is the case.

The kind of Turing machine we've just been talking about we might think of as the common or garden variety. There are lots of different ways to specify this 'normal' type, many of which turn out to be equivalent and some that don't. This is the first type that Turing considered. He then went on to describe a special computational model that's now referred to as a *universal* Turing machine. In this, the tape has two distinct parts. One part encodes the program (as a string of 1s and 0s) while the other holds the input data. The read/write head of a universal Turing machine moves between these parts, carries out the program's instructions on the input, and writes down the output. That's all there is to it: an infinitely long tape that holds *both* the program to be run and the input/output, and a read/write head. A universal Turing machine can perform just six basic operations:

read, write, move left, move right, change state, and stop. Yet, despite this simplicity, it's astonishingly capable. By 'capable' we mean its potential to compute – not superficial qualities such as speed or ease of use.

The fact is, a universal Turing machine, primitive though it may appear, can do everything that any real computer in existence – including any laptop, desktop, mobile device, or large-scale computer – can do. With its simple (though, in theory, infinitely long) tape and read/write head it can replicate every computation possible on the world's most powerful supercomputer. Moreover, it can match anything which any conceivable computer in the future, including such exotica as quantum computers, will be able to do. Quite simply, it can run any possible program.

This may seem surprising. After all, computers in the real world vary enormously. To take an obvious example, different makes of computer run different operating systems such as Windows, Android, macOS, and Linux. Each of these systems has its own unique features and user interface. From a mathematical standpoint, however, they're all the same. All, in fact, are equivalent to a universal Turing machine.

This equivalence leads to the concept of emulation. One computer can emulate, or precisely mimic, another if it can run a program that, from an operational point of view, effectively turns it into that computer. For example, a computer running Windows can execute a program that makes it behave as if it were running macOS. It may not do it very efficiently because it would use a lot of memory and call for a lot of processing, but it could do it. By the same token, fed with the right program any computer, if we suppose it to have an unlimited amount of memory, can be made to emulate any specific Turing machine – including a universal Turing machine. The bottom line is that all real computers and operating systems are mathematically equivalent to a universal Turing machine and therefore to each other, assuming no memory restrictions.

Now, as we've said, Turing's purpose in coming up with his theoretical machine – his general model of computation – had nothing to do with building an actual computer. His goal was to solve Hilbert's decision problem. A universal Turing machine may or may not stop given a specific input. We've already looked at an example of each possibility: where a Turing machine adds a '1' to an existing, finite string of 1s and then stops, and where it writes 1s forever. In these cases it's obvious, ahead of time, what will happen.

But Turing's question was this: is it *always* possible, in the case of any mathematical problem, to determine in advance whether the computation will ever come to an end? For obvious reasons, this became known as the halting problem. Of course, if you weren't sure if a certain program would ever terminate, you could just let it run and run and see what happened. But if it went on for a long time and you chose to give up at a certain point, you'd never know whether the Turing machine was going to stop right after that point or later on, or whether it would carry on forever. Doing case-by-case evaluations like this would prove nothing. The genius of Turing was to describe his computing machine in precise mathematical terms and then ask if there was a general algorithm that could decide the outcome – whether the machine stops or not – for all inputs. In 1936, in a paper titled 'On Computable Numbers with an Application to the *Entscheidungsproblem*', he proved that no such algorithm exists. He then went on to show, in the final part of his paper, that this implies that the *Entscheidungsproblem* can't be solved.

In fact, he wasn't the first to do this. A month before Turing's landmark paper appeared in print, American logician Alonzo Church independently published a paper that reached the same conclusion but using a completely different approach called lambda calculus. Because both Church and Turing showed, around the same time, that a general solution to the *Entscheidungsproblem* is impossible, their result is often referred to as the Church–Turing thesis. There are various ways of expressing this. The gist of it, however, is that it's possible to calculate or evaluate something only if it's computable by a Turing machine, or a device that's equivalent to a Turing machine. This is an important conclusion when it comes to our particular quest for bigger and bigger numbers. It means that although there may be many enormous numbers that are computable by the ultimate means of computation (a Turing machine), there are also numbers that are non-computable.

How can there be numbers, which we can define and know to exist, but that are unreachable by any method of counting or computation – now or in the future? Before we can answer that, we need to finish exploring what's computable, which means staying a while longer within the realm of the Turing machine.

We've said that ordinary computers, such as the laptops on which this book is being written, are Turing machines in every way that matters mathematically. But there are also other things that are equivalent to Turing machines but that bear no superficial resemblance to them because they don't seem to be doing any computations at all. One of these Turing-machines-in-disguise is the Game of Life, devised by John Conway, whom we've already encountered in connection with chained-arrow notation.

The Game of Life wasn't a random idea that popped into Conway's mind one day but, instead, was a development of work first done by Hungarian-American mathematician and physicist John von Neumann and Polish scientist Stanislaw Ulam. We'll be hearing more about von Neumann later because of his pivotal contributions to exploring the limits of maths and what it's possible to compute. As early as 1940, von Neumann was thinking about how the fundamentals of life-as-we-know-it might be recreated inorganically and possibly simulated by a Turing machine. Ulam came up with cellular automata as a way of experimenting with von Neumann's computational theories of life. A cellular automaton is just a grid of cells, each of which can be in a finite number of states. In theory, it could extend in two, three, or more dimensions and involve many different possible states. But the easiest cellular automata to deal with consist of just a 2D array of squares and two possible states per square: 'on' or 'off'. The starting pattern is then allowed to evolve according to a set of rules that's decided in advance.

Von Neumann himself got involved with designing cellular automata that might one day serve as the basis for artificial forms of life built from electromagnetic components. He was particularly interested in whether a hypothetical machine could be contrived that could make exact copies of itself. Von Neumann found that it could by creating a mathematical model for such a machine using very complicated rules on a rectangular grid. But at the time, he was also busy with a lot of other work, including the Manhattan Project, so his efforts in this direction were incomplete. Three decades later, Conway wondered if there was an easier way of proving the same result – the ability to self-replicate – and hit upon the incredibly simple yet fascinating cellular automaton that he called Life.

The universe of Conway's Life is (in theory) an infinite two-dimensional grid of square cells, each of which can be either 'dead' or 'alive'. You can easily play a limited version of it on a sheet of paper, marked into squares, with counters to represent squares that are alive. A game starts out with a particular pattern of living cells and then proceeds in discrete time steps. At each step the new state of every cell is determined by the existing state of its eight immediate neighbours. The rules are straightforward: a living cell with fewer than two live neighbours or more than three live neighbours dies; a living cell with two or three live neighbours survives; a dead cell with exactly three live neighbours comes alive. Although simple, these rules were carefully chosen by Conway so that patterns of cells tend to evolve in interesting and unpredictable ways, neither growing explosively fast nor dying out too quickly.

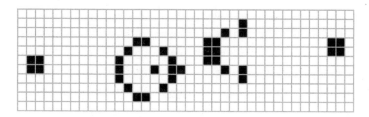

FIGURE 8.2: A Gosper glider gun from Conway's Game of Life.

Conway's remarkable game was first brought to the attention of the wider world through Martin Gardner's Mathematical Games column in the October 1970 edition of *Scientific American*. Gardner introduced his readers to some of the basic patterns in Life, such as the 'block', a single 2-by-2 black rectangle, which under the rules of the game never changes, and the 'blinker', a 1-by-3 black rectangle, which alternates between two states, one horizontal and the other vertical, keeping a fixed centre. The 'glider' is a five-unit shape that moves diagonally by a distance of one square every four turns.

Conway originally thought that no starting pattern would grow indefinitely – that all patterns would eventually reach some stable or oscillating state, or die out altogether. In Gardner's 1970 article on the game, Conway issued a challenge with a $50 reward for the first person who could either prove or disprove this conjecture. Within weeks the prize had been claimed by a team from the Massachusetts Institute of Technology led by mathematician and programmer Bill Gosper, one of the founders of the hacker community. The so-called Gosper glider gun, as part of its endless, repetitive activity, spits out a steady stream of gliders at a rate of one per 30 time steps, or 'generations'.

Combinations of Gosper glider guns, it turns out, can simulate the logic gates that form the basis of computers. A stream of gliders issuing from one of these guns can represent a 'high' signal, or a '1' in binary arithmetic, and an absence of gliders a 'low' signal, or a '0'. One glider can block another because if two gliders meet in the right way, they annihilate each other. There's also something called an 'eater', which is a simple configuration of seven live cells. An eater can absorb excess gliders, thereby preventing them from disrupting other parts of the pattern, while itself remaining unchanged.

Certain configurations of Gosper glider guns and eaters are all it takes to simulate the basic logic gates at the heart of a general-purpose computer or, more to the point, the computational ability of a universal Turing machine. There's nothing that even the world's most powerful, multimillion-pound supercomputer can do that, given enough time and ingenuity, can't be computed using the Game of Life. Furthermore, because Life can be set up to perform as a universal Turing machine it's also impossible to write a program that can predict the ultimate fate of any arbitrary Life pattern, as such a program would then be able to solve the halting problem.

As Turing and Church proved, there's no way of deter-mining in advance, in every circumstance, whether a given program will halt or not. However, this raises another ques-tion. Can we *limit* what a program is capable of in order to guarantee that it terminates? In most programming lan-guages, there's one way to do this easily: by ensuring that the program has no loops. Then, as the program runs from start to finish, it's guaranteed to stop at the end. However, in our particular quest, for larger and larger numbers, this restriction is extremely limiting. Even basic tasks, such as raising a number to a power, generally require some sort of loop to perform a repetitive action like multiplication efficiently. You could try copying the same line of code over and over again to multiply repeatedly, and doing so may allow you to reach the googol. But it becomes a hopeless mission, certainly in practical terms, to get to the googolplex this way – not to mention all the numbers that are far greater still.

Is there then a better way, which would let us surpass the googolplex without allowing programs that may never end? As we saw in the last chapter, one option is to allow only for loops. These may take the form, for instance, 'carry out a series of instructions for n from 1 to 100'. Such a loop will repeat exactly 100 times, and each time the value of n will increment by one, starting at 1 and ending at 100. If we only use these for loops in a program and forbid any other type of recursion, it's possible to prove that every such program terminates. The reason for this is that any individual loop can repeat only a finite amount of times. So, in the case above, you could replace the for loop with 100 copies of its code, with n replaced by the appropriate number in each one. Sometimes a for loop might say 'for n from 1 to m', where m is some other number that the program has previously calculated. In this case, the program would run up to that point in order to figure out exactly how many copies are needed.

If only for loops are permitted, then, as saw earlier, the functions we can compute are known as primitive recursive. Primitive recursive functions are a lot more powerful than non-recursive functions. For instance, it's easy to compute the googolplex in this way (first multiply 10 by itself 100 times to get the googol, then multiply 10 by itself a googol times to get the googolplex). But their power doesn't end there. If you have a function that accepts some input, it's possible to feed that input back into itself, arbitrarily many times. So we can build the function of exponentiation with a for loop, feeding the result of a multiplication back into itself. But the true power becomes apparent when you notice that the exponent becomes the number of times the for loop repeats. We can again feed this back into itself with another for loop, resulting in each repeat of the outer loop creating a massive increase in the number of times the inner loop repeats. We can first calculate the googolplex, then raise 10 to the googolplexth power, then raise 10 to that unimaginable power, and so on, a googolplex times. This brings us to tetration. Using a third for loop gives us pentation, and so on. $3\wedge\wedge\wedge\wedge3$ can be calculated using only a primitive recursive function.

But there are limits to what you can do with primitive recursive functions. You may try to reach Graham's number, but you will inevitably fail. You won't even get close – even g_2 will be beyond your grasp.

So clearly, if you want to make tremendously large numbers and surpass Graham's number, primitive recursion is not the way. So far, we've tried to limit our programs to avoid any possibility of an endless loop. But what if we used the full capacity of our programs? If we allowed any program, no matter how complicated its loops, provided it never loops forever, is there a limit to how large the numbers can get? It turns out that there is still a fundamental limit, going back to the halting problem itself. Before we can understand this limit, however, we have to venture into a very strange land indeed. We have to enter a realm beyond the finite.

CHAPTER 9

Infinite Matters

NOTHING LASTS FOREVER – except in infinite time. Nothing goes on forever – except in infinite space. No finite number is as big as an infinitely large number. If we're going to allow infinity to muscle its way into our quest to find the biggest number in the world, that may seem like cheating because all the numbers we've talked about so far, however vast they may be, are finite – they all lie at specific points on the real number line. How could a finite number, however big it is, compete with something that's infinitely big? For the moment, though, that's not the point. Strange to say, in order to reach finite numbers of even greater size than those we've so far encountered we need to enlist the help of the infinite.

Infinity is a slippery concept. We can't properly grasp it in our minds, because our minds are finite, capable only of finite thoughts, yet we recognise that infinity might exist in the real world. Cosmologists debate whether the universe is limited in time and space or whether it goes on forever. Current observations suggest the latter: that it may be infinite in extent and duration. But we don't yet have enough information to know for sure. We also don't know what happened before the Big Bang – the instant at which all the contents of the universe began their outward rush from a state of incredibly high density and pressure. *Was* there a before? And if there wasn't, how could the universe have made the transition from a state of non-existence, outside of time, to something that was tangible and temporal in nature?

Physical infinities, in the form of infinitely large spaces and times, or infinitely high densities and temperatures, are concepts we can discuss but never truly grasp. The same is true of the infinite in mathematics. We can talk about the number line going on forever without reaching an end yet have no meaningful sense of what an infinitely long number line would be like. Until the latter part of the nineteenth century, philosophers and mathematicians drew a sharp distinction between what Aristotle, a couple of thousand years earlier, had called 'potential' and 'actual' infinity. There is only an 'illimitable potentiality of addition', he said, and an 'illimitable potentiality of division', for 'it will always be possible to find something beyond the total.'

Most mathematicians were happy to accept the notion of potential infinity. In terms of the number line, for instance, they recognised that there was no point at which it abruptly stopped and that, however far you travelled along it, there would always be more numbers beyond. By the same token, it was known that irrational numbers, such as pi and the square root of 2, had decimal expansions that never ended nor repeated in any predictable pattern. But there was still a reluctance to accept infinity in its entirety – as a fully formed thing in itself and not merely an interminable extension of the finite.

Mathematicians refused, or simply didn't know how, to deal with actual infinity. They were aware of it hovering over their subject like a threatening storm cloud but turned a blind eye to it. Even Carl Gauss, colossus of mathematics that he was, expressed in 1831 his 'horror of the actual infinitude':

> I protest against the use of infinite magnitude as something completed, which is never permissible in mathematics. Infinity is merely a way of speaking, the true meaning being a limit which certain ratios approach indefinitely close, while others are permitted to increase without restriction.

This state of denial didn't prevent mathematicians from developing key concepts, such as infinite series, limits, and infinitesimals, by, for instance, letting '*n* tend to infinity'. Isaac Newton and Gottfried Leibniz, independently, were able to lay the foundations of calculus without having to grant that infinity was a new type of mathematical object. But, in retrospect, it's clear that something eventually had to give. Even by the dawn of the seventeenth century, certain paradoxes and puzzles had arisen, which suggested that actual infinity wasn't an issue that could be brushed aside forever. These conundrums stemmed from the observation that it's possible to pair off all the members of one collection of objects with all those of another of equal size, leaving none left over. Applied to indefinitely large collections, this pairing-off principle seemed to fly in the face of a common-sense idea first expressed by Euclid, namely, that the whole is always greater than any of its parts. For instance, it appeared that all the positive integers could be paired off with only those that are even: 1 with 2, 2 with 4, 3 with 6, and so on, despite the fact that positive integers also include odd numbers. Galileo, in considering such a problem, was the first to show a more enlightened attitude towards the infinite when he said: 'Infinity should obey a different arithmetic than finite numbers.'

But what kind of 'different' arithmetic could it be? And how could something as bizarre as infinity – endlessness fully realised in the moment – have a place in mathematics? Potential infinity, regarded as the limit of an eternal quest, was by contrast a comfortable notion and, as Aristotle pointed out, was reflected in the unending cycle of the seasons or the apparent limitless divisibility of a nugget of gold (the existence of atoms being then unknown).

The concept of potential infinity lulls us into thinking that if we just keep going far enough, or long enough, we'll get closer to infinity. It's then only a small step to the popular myth that infinity is just like a very big number, and that a trillion, or a trillion trillion trillion, is somehow closer to infinity than, say, ten is or a thousand. But this simply isn't the case. Travelling further down the number line or counting to bigger and bigger numbers gets us *no nearer whatsoever* to infinity. We're just as far from, or close to, infinity at the number 1 as we are at any finite number we care to name, however vast. Put another way, all of infinity is contained between any two numbers, no matter how close they are, so that setting off down the road to larger and larger numbers in search of infinity is utterly futile. The fact is that infinity exists between 0 and 1, for instance, because there are infinitely many fractions – ½, ⅓, ¼, and so on. Infinity is not like a big finite number at all. To deal with infinity we have to jump out of the realm of finite numbers altogether and stop using them as a crutch to our understanding.

The German mathematician David Hilbert conjured up a striking mental picture to illustrate how strange the arithmetic of infinity can be. Imagine, he said in a lecture given in 1924, a hotel with an infinite number of rooms. Unlike an ordinary hotel, which can't accommodate any more guests once all its rooms are occupied, 'Hilbert's Grand Hotel' can always find more space. If a new guest arrives, all the manager has to do is ask everyone currently in residence to move into the next higher room number. The guest in room 1 relocates to room 2, the occupant of room 2 to room 3, and so down the corridors and floors that stretch away forever, so that the newcomer can be placed in room 1. A thousand new guests could arrive and not be disappointed – all current residents would simply be asked to switch to the room number that was 1,000 higher. No one need ever be turned away. Even if an *infinite* number of people suddenly arrived, unannounced, every single one could be accommodated by moving the occupants of rooms 1, 2, 3, and so on, to rooms 2, 4, 6, etc., thus freeing up all the (infinitely many) odd-numbered rooms. This most capacious of hostelries need never turn anyone away – even if an infinite number of coaches were to arrive one morning, each carrying an infinite number of passengers.

Hilbert's Hotel is a weird and wonderful place that doesn't fit comfortably with any of our common-sense notions of how the world works. But common sense isn't much use in dealing with the infinite. The fact is, properties of infinitely many things are fundamentally different from those of finitely many things. It may be hard to swallow, but when it comes to Hilbert's Hotel, the statements 'there's a guest in every room' and 'more guests can be accommodated' aren't mutually exclusive.

Whenever a revolution starts to brew, whether in science or mathematics, intellectual battle lines are drawn between the old guard and the new. Controversy erupted over Einstein's theories of relativity and, slightly earlier, at the dawn of the twentieth century, when the suggestion began to circulate that energy might be quantised. In the late nineteenth century, mathematicians faced a crisis of their own: Were they ready to accept that actual infinity was a mathematical object? The majority, aligning themselves behind Aristotle, Gauss, and other greats who in the past would tolerate only potential infinity, were not. Only a handful of theorists were prepared to wave the banner for change: to build the foundations of an entirely new branch of mathematics – set theory.

At the forefront of this revolution were three German mathematicians: Richard Dedekind, Karl Weierstrass, and, head and shoulders above the others, Georg Cantor. Set theory and, with it the concept of infinite sets, burst upon the mathematical scene with the publication of a single paper by Cantor in 1874 titled 'On a Property of the Collection of All Real Algebraic Numbers'.

Cantor realised that the age-old pairing-off method used to tell if two finite sets are equal, could be applied just as well to infinite sets. It followed that there really are just as many even positive integers as there are positive integers altogether. Far from being a paradox, Cantor saw that this, in fact, was a *defining property* of infinite sets: that the whole is no bigger than some of its parts. He went on to show that the set of all natural numbers, **N**, which is the set of all non-negative integers, 0, 1, 2, 3, … (sometimes 0 is not included), contains precisely as many members as the set of all rational numbers, **Q** – numbers that can be written in the form of one whole number divided by another. He called this infinite number aleph-null ($\aleph_0$), 'aleph' being the first letter of the Hebrew alphabet and 'null' the German for 'zero'. It's also known as aleph-zero or aleph-nought.

You might be wondering why call it 'aleph-null' and not just 'aleph', because surely there can be only one kind of infinity? Well, first of all, aleph-null isn't infinity because infinity is a broad philosophical concept not a precise mathematical thing. Aleph-null is the size of the infinite set of natural numbers. It's what's referred to as a transfinite number because it transcends the size of any finite number. Second, surprisingly, some infinite sets *are* bigger than aleph-null. In fact, aleph-null is the smallest transfinite number – even though it's infinitely large!

How can this be? The natural numbers go on forever. What could it possibly mean to say that some sets of numbers go on forever – and then some? Again, it's important to be mathematically precise and not try to conjure up some vague mental image of what different types of infinity might be like. Cantor was able to show that the size of the sets of all natural numbers, and all integers, and all rational numbers was the same and equal to aleph-null. Bigger than aleph-null is aleph-one, which we'll define later in the chapter. According to what's known as the continuum hypothesis, aleph-one is the size of the set of all real numbers, **R**, which includes both rational and irrational numbers. The continuum hypothesis may or may not be true depending on which version of set theory you're working with (something else we'll come back to later!). But, for now, thinking of aleph-one as the size of the real number set is useful because it illustrates how one transfinite number can be bigger than another. On the real number line, irrational numbers, such as $\sqrt{2}$, are infinitely more dense (or common) than are rational numbers, so that the set of all real numbers is infinitely larger than the set of rational or natural numbers.

The hierarchy of alephs doesn't stop at aleph-one. There's also aleph-two, aleph-three, and so on, each infinitely bigger than the one before. In fact, there are infinitely many different alephs. Not only that, but corresponding to each aleph, it turns out, are infinitely many other infinitely large numbers. To understand this, we have to explore the important difference, in the realm of the transfinite numbers, between cardinals and ordinals.

In everyday language and arithmetic, cardinal numbers tell us how many there are in a collection of things – one, six, fifty-seven, and so on, whereas ordinal numbers, as the name suggests, give the order or position of something – first, sixth, fifty-seventh, etc. As far as ordinary situations are concerned, then, there's not much difference between cardinals and ordinals. And any difference there is seems pretty obvious.

Say we're talking about matches. It's clear that you can't have a sixth match without having at least six matches in a group, and that you could still have a sixth match even if there were eight in a group. You could also have six matches without having a sixth one if you didn't put them in any particular order – say by just making a pile of them. But, these minor distinctions aside, we can use the same symbols for cardinals and ordinals – 1 (or 1st), 6 (or 6th), 57 (or 57th), and so on – and not worry too much over how they differ.

When it comes to infinite sets of numbers, however, the distinction between cardinals and ordinals becomes vitally important. To understand why, we need to look at the basics of set theory – the powerful new branch of mathematics that Cantor, along with Dedekind, was instrumental in developing.

Informally, a set in maths is like any other kind of set, such as a collection of stamps or vinyl records – just a grouping of objects. It's described by listing its members, or elements, separated by commas, within braces. For example, {1, 8, 64, 125} and {cat, T, apple, 3} are both sets. The size of a set – how many members, or elements, it contains – is known as its cardinality and is given by a cardinal number. Both the sets just mentioned have four elements and so have a cardinality of four. In general, the cardinality of two sets is the same if every member in the first set can be paired off with one in the second so that nothing is left over; in other words, they have a one-to-one correspondence. For example, we can pair 1 with cat, 8 with T, 64 with apple, and 125 with 3 to show that these sets have the same cardinality. The finite cardinals – the cardinals that measure the size of finite sets – are just the natural numbers 0, 1, 2, 3, and so on.

In the case of finite sets, the difference between the size of a set, given by a cardinal number, and its 'length', given by an ordinal number, is so slight as to be almost pedantic. But when it comes to infinite sets, Cantor realised, these are two very different things. To grasp how different they are we need to understand the idea of a 'well-ordered' set. A set is considered to be well ordered if each subset, or sub-grouping, of its members has a first member (provided, that is, the subset isn't empty). The finite set {0, 1, 2, 3}, for instance, is well ordered. The set of all integers, on the other hand, which includes all negative whole numbers as well as all positive ones, {...−2, −1, 0, 1, 2, ...}, isn't well ordered because there's no first member. The set of all natural numbers, {0, 1, 2, 3, ...} is well ordered because despite having no specified member at the end it has one at the start and every subset containing only natural numbers also has a first member.

Now, a key point is that well-ordered infinite sets of the same size, or cardinality, *can have different lengths*. That isn't an easy concept to grasp, even for a mathematician. Strictly speaking, we should say different 'ordinalities' rather than 'lengths', but the more familiar term helps us to appreciate what's going on. Think about the well-ordered sets {0, 1, 2, 3, 4, ...} and {0, 1, 2, 4, ... 3}, where the three dots mean 'carry on forever'. Both sets contain all the natural numbers and therefore have the same size or cardinality, aleph-null. But the second is slightly longer. At first, this doesn't seem to make sense. After all, if we were talking about finite sets then it's obvious that {0, 1, 2, 3, 4} and {0, 1, 2, 4, 3} are identical in length because they both contain five members. But infinite sets are fiendishly counterintuitive. The set {0, 1, 2, 3, 4, ...} has no finite end member because the three dots tell you to carry on forever without stopping. However, {0, 1, 2, 4, ... 3} is different. It too contains a sequence of members that carries on forever. However, it also contains one member that is beyond all the members of the never-ending sequence and that's been detached from it. With the 3 taken out, the sequence 0, 1,

2, 3, … is just as long as 0, 1, 2, 4, …; in other words, you could pair off all the members of these two sequences and never have one left over. But moving the 3 to the end, so that it comes after the infinite sequence, adds one to the length. Think of it another way. With the first set, $\{0, 1, 2, 3, 4, …\}$, there's a first element (0), a second element (1), a third element (2), a fourth element (3), and so on. With the second, there's also a first (0), second (1), third (2), fourth (4), etc. However, there's one element, 3, that is none of these. The ordinal we assign to 3 – not the value of the number but the order in which it appears – is greater than anything that comes before it.

We need a naming system for this class of infinite numbers, which is different from the alephs. Mathematicians call the smallest infinite ordinal – the shortest length of the set of all natural numbers – 'omega' (ω). The ordinality of the set, $\{0, 1, 2, 4, … 3\}$, where the 3 is placed after all the other natural numbers, is one greater, namely, $\omega + 1$. Another way of saying this is that 3 is the $(\omega+1)$th element in the set $\{0, 1, 2, 4, … 3\}$. The '+' sign here is a bit misleading because it doesn't mean addition in the usual sense but, rather, that $\omega + 1$ is the next ordinal after ω. Ordinal subtraction also works very differently from the way we're used to taking away. The ordinality of $\{0, 1, 2, 4, …\}$, with the 3 taken out, is still ω. For instance, if we want to compute $\omega - 3$, we can't simply remove elements from the end of the set of all natural numbers, since there isn't an end. Instead, we remove elements from the start, but in doing so we find that $\omega - 3$ is just ω again! The fact is that the 'length' of the set of all natural numbers can't be reduced, no matter how large a finite number of elements is removed from it. On the other hand, its length *can* be increased by putting the elements that have been removed at the end.

To recap: aleph-null and ω both refer to the same set – the set of natural numbers. Aleph-null is its size (how many elements it contains) and ω is its shortest length. This length can be increased by taking elements out of their usual order and placing them at the end. The set $\{2, 3, 4, \ldots 0, 1\}$, for instance, has a cardinality of aleph-null but an ordinality of $\omega + 2$. We can keep on increasing the length of the set of natural numbers by moving more and more elements beyond the three dots that mean 'carry on forever': $\omega + 3, \omega + 4, \ldots,$ all the way up to $\omega + \omega$ (or $\omega \times 2$), which could be written, for instance, as the subset of all even numbers followed by the subset of all odd numbers, $\{0, 2, 4, \ldots, 1, 3, 5, \ldots\}$, since each of these is equal in length to ω. Then we can continue, as before, by shifting elements to the end; for instance, one way to write $\omega \times 2 + 1$ is $\{2, 4, \ldots, 1, 3, 5, \ldots, 0\}$. Then we can move on to powers of ω, such as ω^2, ω^3, $\ldots$, all the way up to ω to the power of ω (ω^ω), and then to stacks of powers (power towers) of ω, stretching higher and higher until we reach a power tower of ωs that is ω high. At this point we reach a new level – an ordinal that Cantor called epsilon-zero (ε_0). Just as ω is the smallest ordinal that lies beyond the finite ordinals, ε_0 is the smallest ordinal that lies beyond any ordinal that can be expressed in terms of ω using addition, multiplication, and exponentiation. It's the gateway to the realm of epsilon numbers, which, like that of the omega ordinals, is infinitely large. The whole process described for the omegas can be repeated for the epsilons until all the mathematical operations that are possible using epsilons, including power towers of epsilons or even epsilons of epsilons, are exhausted. At this point we arrive at yet another level of infinite ordinals, starting with zeta-zero (ζ_0). And so it goes on and on and on.

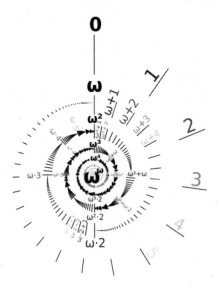

FIGURE 9.1: A representation of the ordinal numbers up to ω^ω. Each turn of the spiral represents one power of ω.

More than anything, the difficulty in progressing further is one of notation. Eventually, all the Greek letters are exhausted, along with any other ordinary labelling system, to represent the hierarchy of infinite ordinals that stretches away into the distance. Compounded with the problem of finding more powerful and compact means of notating vast infinite ordinals is a mounting degree of technical difficulty. Some milestones, named after the mathematicians with whose work they're associated, lie along the way, once zeta-zero has been left far, far behind: the Feferman–Schütte ordinal, the small and large Veblen ordinals (both of which are outrageously large), the Bachmann–Howard ordinal, and the Church–Kleene ordinal (first described by the American mathematician Alonzo Church and his student Stephen Kleene). To describe properly what any of these mean would take a book in itself, so esoteric is the maths behind them. The Church–Kleene ordinal, for instance, is so incomprehensibly vast that there's *no* notation whatsoever that can reach up to it.

These ordinals are rarely encountered by professional mathematicians, let alone the wider public, but the essential point about them all is that they're *countable*. In other words, all the infinite ordinals we've talked about so far, starting with ω, can be paired off one-to-one, leaving none left over, with the natural numbers. Another way of saying this is that they all correspond to the cardinality aleph-null. We're no nearer to a bigger kind of infinity when we get to epsilon-zero or even the mighty Church–Kleene ordinal than when we started: colossal though they may be, they merely represent different ways of ordering the set of all natural numbers. A bigger kind of infinity means one that transcends aleph-null altogether. But how is that possible?

Aleph-null doesn't behave like the numbers we're used to dealing with. Whereas $1 + 1 = 2$, aleph-null + 1 is still aleph-null. Aleph-null plus any finite number or minus any finite number is still aleph-null. This suggests a new twist to the song 'Ten Green Bottles', along the lines of: 'Aleph-null green bottles hanging on the wall, aleph-null green bottles hanging on the wall, and if one green bottle should accidentally fall, there'd be aleph-null green bottles hanging on the wall' (repeat *ad infinitum*). You can't change aleph-null by subtracting from it, adding to it, or multiplying it by any finite number or even multiplying it by aleph-null itself. But Cantor showed, using a theorem that's now named after him, that there is a hierarchy of infinities of which aleph-null is the smallest. The next infinite cardinal, aleph-one, is much bigger and equal in size to the set of all countable ordinals, namely those with cardinality aleph-null.

Because the natural numbers are countable, aleph-zero, the size of the set of natural numbers, is said to be a countably infinite cardinal. Corresponding with it is the smallest infinite countable ordinal, ω, and infinitely many other countably infinite ordinals. All of these infinitely many countable ordinals arise because, in the case of ordinals, information on order matters so that a much finer distinction has to be made than with cardinals. Even so, all the countable ordinals from ω onwards, including the epsilon numbers and the rest, fall within the same cardinality – aleph-null. But with aleph-one comes a dramatic change. Not only is aleph-one indescribably larger than aleph-null but it's also *uncountable*. Corresponding to it is the smallest uncountable ordinal: omega-one (ω_1).

We've said that aleph-one is the size of the set of countable ordinals but does it have any other interpretation? Aleph-null measures the size of the set of all natural numbers. Does aleph-one also correspond with anything that's familiar and conceptually easy to grasp? Cantor thought so. He believed that aleph-one was identical with the total number of points on a mathematical line, which, astonishingly, he found was the same as the number of points on a plane or in any higher *n*-dimensional space. This infinity of spatial points, known as the *power of the continuum*, *c*, is also the set of all real numbers (all rational numbers plus all irrational numbers). Cantor's continuum hypothesis asserts that c equals aleph-one, which is equivalent to saying that there's no infinite set with a cardinality between that of the natural numbers and that of the real numbers. Yet, despite much effort, Cantor was never able to prove or disprove his continuum hypothesis. We now know why – and it strikes to the very foundations of mathematics.

In the 1930s, Austrian-born logician Kurt Gödel showed that it's impossible to prove the continuum hypothesis is wrong starting out from the standard axioms, or assumptions, of set theory. Three decades later, the American mathematician Paul Cohen showed that neither can it be proved correct from those same axioms. Its status was, in other words, indeterminate within the normal framework that mathematicians used. Such a situation had been on the cards ever since the emergence of a famous theorem, discovered by Gödel, called the incompleteness theorem. But the independence of the continuum hypothesis was still unsettling because it was the first concrete example of an important question that provably couldn't be decided either way from the universally accepted system of axioms on which most of mathematics is built.

The debate about whether the continuum hypothesis is ultimately true or not, or whether it's even a meaningful statement, rumbles on among mathematicians and philosophers. As for the nature of the various types of infinities and the very existence of infinite sets, these depend crucially on what number theory is being used. Different axioms and rules lead to different answers to the question *what lies beyond all the integers*? This can make it difficult or even meaningless to compare the various types of infinities that arise and to determine their relative size, although within any given number system the infinities can usually be put into a clear order.

There is a towering hierarchy of cardinals beyond aleph-null. Assuming the continuum hypothesis to be true, which is the default position of most mathematicians (because it has helpful consequences), the next cardinal is aleph-one, equal to the size of the set of all real numbers. After this comes aleph-two, then aleph-three, and so on, without end. To each aleph corresponds an infinite number of ordinals, the smallest of which is ω in the case of aleph-null, ω_1 in the case of aleph-one, ω_2 in the case of aleph-two, and so on.

Are any of these mathematical infinities enacted in the real world or are they pure abstractions? We saw earlier that cosmologists are leaning towards the view that the universe we live in is geometrically flat and endless in space and time. If it does go on forever, with which kind of mathematical infinity does it correspond? The fact that space and time appear to come in discrete amounts – the Planck length and Planck time – means that they're not continuous like the points on a mathematical line. So, if the actual universe is infinitely large it seems that it could correspond with only the smallest kind of infinity, aleph-null. Anything bigger may always be confined to our intellects or some Platonic space unfettered by the laws of physics.

CHAPTER 10

Growing Fast

COUNTING 1, 2, 3, and so on, would eventually get us to the biggest finite numbers we've talked about so far. But there are two obvious problems with travelling down the number line in this leisurely way if our goal is to study and name numerical giants. First, counting up one at a time is incredibly slow. And second, the numbers you'd eventually reach (after many times the present age of the universe!), such as the googolplex, are so big that they can't be written in full as a decimal numeral.

The advantage of realising that it's possible, if only in principle, to get to colossal numbers simply by counting is that it reminds us that the only thing that's different about such numbers *is* their size. They aren't different in nature from smaller numbers like 3, 28, or 1,016, any more than the distance across your hand is different in nature from the distance to some far-off galaxy. It's just a matter of scale. But, as we've already discovered, the problem when dealing with seriously large numbers is devising ways to represent them that are compact, meaningful, and mathematically accurate. In the case of Graham's number, for instance, we know that it exists, in some sense, as a long string of digits, the last of which are 262464195387. It's precisely defined and has a unique value. But there's no way actually to write it in full or in any other form, such as an exponential, using maths that we're normally taught in school. That's why, earlier in this book, we started to bring in some new ideas, such as the hyperoperator series, so that we had more powerful tools to take us into big-number territory. Now, we're going to go further, to the very limit of what's computable. To do this we'll be hitching a ride on something called a fast-growing hierarchy.

If we go back to the analogy with propulsion systems that can hurl us across vast distances at blindingly high speeds, a fast-growing hierarchy is the equivalent of warp drive. Its mathematical technology hinges on three key concepts: functions, ordinals, and recursion. We've already come across each of these in various guises but it'll be good to recap to make sure we're starting on firm ground.

A function in maths, remember, is just a relationship, or a rule, for turning inputs into outputs. So a function is like a mechanism that transforms one set of values into another set by always going through the same process. The process might be, for instance, 'add 3 to the input'. If we call the input x and the function $f(x)$, then $f(x) = x + 3$. We can also have functions of functions. For instance, $f(f(x))$ feeds in $f(x)$ as the input, so that $f(f(x)) = (x + 3) + 3 = x + 6$. And if we can have functions of functions, then why not functions of functions of functions, and so on? Continuing with our example, and working right to left, $f(f(f(x)))$ feeds $f(f(x))$ into $f(x)$ to give $((x + 6) + 3) = x + 9$.

An ordinal, as we saw in the last chapter, describes a way

of arranging a collection of numbers (or other objects) in a specific way, so that there's a first, a second, a third, and so on. If the collection is finite it can be put in order just by counting – by labelling its members with distinct natural numbers. But there are also infinitely large ordinals, the smallest of which is ω (omega) – the shortest ordering of a set that has aleph-null members. In a fast-growing hierarchy, ordinals are used to 'index' the functions, which means that each function in the hierarchy is labelled with an ordinal.

Let's start nice and easy with a function that just adds one to a number and call it f_0. So, if the number we want to apply the function to is n, then $f_0(n) = n + 1$. You may recognise this as the 'successor' function we met earlier – a function that advances us down the number line to the next natural number. As you know, this isn't going to get us to really big numbers any time soon – it's just counting up in steps of 1. So let's move on to $f_1(n)$. We've taken our first step up the fast-growing hierarchy, increasing the ordinal index – the subscript after the f – from 0 to 1. This new function feeds the previous one into itself n times, in other words $f_1(n) = f_0(f_0(\dots f_0(n))) = n + 1 + 1 + 1 \dots + 1$, where there are n ones, giving a total of $2n$. Again, this isn't too impressive in terms of how quickly it can get us into the land of giant numbers. But it reveals the process that ultimately gives a fast-growing hierarchy its immense power: recursion.

The next rung up the ladder, $f_2(n)$, feeds $f_1(n)$ into itself n times. So, we can write $f_2(n) = f_1(f_1(\dots f_1(n))) = n \times 2 \times 2 \times 2 \dots \times 2$ with n twos. This is the same as $n \times 2^n$, which is an exponential function. Plugging in a value of, say, 100 for n, we get $f_2(100) = 100 \times 2^{100} = 126,765,060,022,822,940,149,670,320,537,600$, or about 127 billion trillion trillion.

That's quite a jump. If $n = 100$, we've gone from of $f_0(100) = 100 + 1 = 101$, to $f_1(100) = 100 + (1 \times 100) = 200$, to a number in the billions of trillions of trillions. And that's just the start.

The function $f_3(n)$ involves n repetitions of $f_2(n)$ and leads to a number slightly greater than 2 to the power 2 to the 2 to the 2... where the power tower is n high. This brings us to the stage of two up-arrows, or tetration – the operation we came across earlier when making our assault on Graham's number. If $n = 100$, then $f_3(n)$ has already gone way past the capacity of the universe to contain it if we tried to write out the result in full.

Continuing in the same vein, $f_4(n)$ involves three up-arrows, $f_5(n)$ four up-arrows, and so on, every increase of the index ordinal by one having the effect of adding one more up-arrow and boosting the number of up-arrows to $n - 1$. This certainly takes us into big-number territory by everyday standards but it amounts to nothing more than the hyperoperator series that we've already seen. Just adding one up-arrow at a time would never get us to Graham's number, let alone anything vastly bigger, in a sensible amount of time. A new approach is needed. To reach truly colossal finite numbers, we need to transcend everything that's come before: we must enlist the help of numbers that are infinitely large. *This* is where a fast-growing hierarchy really comes into its own: when it taps the power of infinitely large ordinals to control how many times a function has to be performed. But before we can properly wrap our brains around this idea, we need to delve a bit more into the mathematics of transfinite, or infinitely large, numbers – in particular, the maths of transfinite ordinals.

The smallest kind of infinity, as we saw in the last chapter, is aleph-null, the size of the set of all natural numbers. The mathematics of it, we discovered, is both alien and counterintuitive. Add or take away any finite number from aleph-null and the result is still aleph-null. For example, $\aleph_0 + 1 = \aleph_0$ and $\aleph_0 - 300 = \aleph_0$. Multiplication and division by finite numbers, similarly, has no effect: $\aleph_0$ times a million is still $\aleph_0$. You can even add aleph-null to itself and the result stubbornly remains aleph-null.

Aleph-null is a transfinite cardinal. Cardinals measure how many members a set contains. In the case of aleph-null, the set in question is that of natural numbers: $\{0, 1, 2, 3, \ldots\}$. Although this has a fixed number of members, these members can be rearranged so that the set has different ordinalities, or 'lengths'. The shortest length of the set of all natural numbers is the smallest transfinite ordinal: omega (ω). The next shortest is $\omega + 1$, then $\omega + 2$, then $\omega + 3$, and on and on. A key feature of all these different transfinite ordinals, which correspond to different ways of ordering the set of all natural numbers, is that they're *countable*. This doesn't mean that you could actually count them all but rather that there's a one-to-one correspondence between the ordinal and the natural numbers.

Now, before we move on and see how all this applies to our fast-growing hierarchy, we need to know about a few other concepts to do with ordinal numbers. The first of these are the concepts of 'successor ordinal' and 'limit ordinal'. A successor ordinal is exactly what its name suggests: any ordinal that immediately succeeds, or comes after, some other ordinal. The easiest way to think of this is that every successor ordinal is one greater than the ordinal that precedes it. All natural numbers that are greater than zero are successor ordinals. A limit ordinal is a nonzero limit of a sequence of ordinals, all of which are less than it, without itself being a successor. So, every ordinal must be one of the following: zero, a successor ordinal, or a limit ordinal.

You can't reach a limit ordinal by adding 1, or any other finite number, to any ordinal that is less than it. Following this definition, the smallest limit ordinal is ω. The next smallest ordinal after ω is $\omega + 1$, but this is a successor ordinal. Infinitely many more successor ordinals follow until we reach the next limit ordinal: $\omega + \omega$. This is a limit ordinal for the same reason that ω is: namely, it isn't the result of adding 1, or some other finite number, to any number that it exceeds in size. Corresponding to each limit ordinal is what's called a fundamental sequence – a sequence of ordinals that approach the limit ordinal (but never reach it) from below. In the case of ω, the fundamental sequence is just the sequence of natural numbers: 0, 1, 2, 3, ... Any ordinal built using ω has a fundamental sequence based on this. So, for instance, the fundamental sequence of the limit ordinal $\omega*2, = \omega + \omega$, is $\omega, \omega + 1, \omega + 2, \omega + 3, \ldots$, while ω^2 has the fundamental sequence $\omega, \omega*2, \omega*3$, and so on. In general, the last ω can be replaced by its fundamental sequence (expanding multiplication and exponents as appropriate), so, for example, ω^2 is rewritten as $\omega*\omega$.

At last, we're ready to appreciate the true power of the fast-growing hierarchy, by bringing transfinite ordinals into play. We'll see how the use of these ordinals to index the functions serves as a springboard for reaching some of the largest computable *finite* numbers ever conceived. We started off with $f_0(n), f_1(n), f_2(n)$, and so on, gradually working our way up the hierarchy, using natural numbers as indices. Now we're going to jump to $f_\omega(n)$, the function indexed by the smallest transfinite ordinal, ω. But there's a problem we have to overcome in doing this. When we were figuring out the value of functions indexed by natural numbers, the first step was to reduce the index by one and write down a recursion relation. For example, $f_3(n) = f_2(f_2(\ldots f_2(n)))$. But we can't use this approach when the index is ω because ω is a limit ordinal and therefore isn't the successor of any ordinal.

Instead, what we do is replace ω with the nth member of its fundamental sequence – namely, n itself – so we find that $f_\omega(n)$ is just $f_n(n)$. Now, to be clear, we're not saying that $\omega = n$, despite how it may appear. What we're doing is expressing $f_\omega(n)$ in terms of (finite) ordinals smaller than ω, so that we can reduce the function to a form that's useful for doing calculations. Perhaps you're thinking we may as well write $f_n(n)$ instead of $f_\omega(n)$ and get the same result, but that would prevent the next crucial step – the step at which the full potential of a fast-growing hierarchy becomes apparent.

As soon as we go from $f_\omega(n)$ to $f_{\omega+1}(n)$ something dramatic happens. Remember, when the ordinal that indexes the function goes up by 1 this feeds the previous function back into itself n times. If using a finite ordinal results in a fixed number of up-arrows, and using ω produces $n-1$ up-arrows, then using $\omega + 1$ allows us to feed back into the number of up-arrows n times, which amounts to a fantastic jump in the strength of the recursion.

To understand this, think about the function $f_{\omega+1}(2)$, which equals $f_\omega(f_\omega(2))$ using our recursion rule. Because we defined $f_\omega(2)$ to be the same as $f_2(2)$, we can rewrite $f_{\omega+1}(2)$ as $f_\omega(f_2(2))$, just replacing the innermost ω with 2. (We can't work out the value of the outer f_ω until we know what value the inner one will take.) As it turns out, $f_2(2) = 8$ so now we're left with $f_{\omega+1}(2)$, which equals $f_\omega(8)$. Finally, we can simplify the outermost ω and get $f_8(8)$, which is a number with seven up-arrows. While this shows how $f_{\omega+1}$ can be used to feed back into the number of up-arrows, it doesn't give a clear impression of the function's awesome capability. This only becomes apparent as n gets larger, and the corresponding number of feedback loops grows. Putting $n = 64$ gives $f_{\omega+1}(64)$, which is approximately Graham's number. The next step up the fast-growing hierarchy, $f_{\omega+2}(n)$, bursts into totally new territory because it plugs all of the mathematical machinery used to reach the level of Graham's number back into itself. The result is a number we can write roughly as $g_{g...64}$ – with 64 levels of the g subscript! There's no hope of trying to grasp, even vaguely, what this means but it obviously represents a game-changing explosion in size.

With each step up the hierarchy – $f_{\omega+3}(n)$, $f_{\omega+4}(n)$, …, – there's a further, extraordinarily steep rise in the number of recursive operations acting on what is already, from the previous step, a spectacularly large number. The countably infinite ordinals stretch away into the far distance, each successive one providing the basis for a recursive function that utterly dwarfs the power of the one before it. The omegas alone form a sequence so long that it only ends when we reach omega raised to a power tower of omega omegas. This mighty ordinal, known as epsilon-zero, is so vast that it can't be described within our conventional system of arithmetic, known as Peano arithmetic. With each step along the eternal road of omegas, the finite number generated by recursion jumps by an amount that defies comprehension. But beyond the loftiest omegan power tower, lie tier upon higher tier of yet greater infinite ordinals – first the epsilons, then the zetas, and on and on, without end – as we saw in the last chapter in our exploration of infinity. These increasingly vast ordinals serve to define more and more powerful degrees of feedback.

Finally, we reach a tremendously large ordinal known as Gamma-zero (Γ_0) or, more magnificently, the Feferman–Schütte ordinal, after the American philosopher and logician Solomon Feferman and the German mathematician Kurt Schütte who first defined it. Although Gamma-zero is still countable, and there are countable ordinals beyond it, actually defining it requires the use of uncountable ordinals (ones that can't be made from rearranging elements of aleph-null but instead require aleph-one or more elements). This process is reminiscent of how the fast-growing hierarchy itself evolves. Just as we have to resort to using infinite ordinals in the fast-growing hierarchy to describe huge finite numbers, so we need to turn to uncountable ordinals to describe truly tremendous countably infinite ordinals. There aren't any adjectives left to describe adequately the size of the finite numbers to which the Feferman–Schütte ordinal, and others beyond it, give rise by recursion. Nor does any mathematician have a brain big enough or clever enough to truly grasp the immensity of the numbers their recursive techniques can spawn. Nevertheless, these numbers exist, they're finite in size (and therefore lie on the familiar number line that we learn about as young children), and they could, in principle, be reached by counting up, one by one, from zero.

In our quest to find the biggest number in the world we've alternated between looking at specific examples of large numbers and ways of generating and defining them. Among the former, for instance, are the googol, the moser, and Graham's number. The latter include the hyperoperator sequence, Conway's chain arrows, and now this astoundingly powerful method of using a sequence of functions indexed by an endlessly increasing succession of transfinite ordinals.

We've been talking about *a* fast-growing hierarchy rather than *the* fast-growing hierarchy for a good reason – there are different varieties. These varieties may differ both in the choice of initial function f_0 and in the choice of fundamental sequences at the limit ordinals. The one thing that all fast-growing hierarchies have in common is that they're sequences of rapidly increasing functions indexed by ordinals.

One of the first examples of the use of this concept was provided by G. H. Hardy in 1904. The Hardy hierarchy starts off in a very pedestrian way: $H_0(n) = n$, $H_1(n) = n + 1$, $H_2(n) = n + 2$, and so on, following the same pattern for all finite ordinals. Even when it gets to transfinite indices, its pace isn't exactly blistering: $H_\omega(n) = 2n$, $H_{\omega+1}(n) = 2n + 2$, $H_{\omega+2}(n) = 2(n + 2)$, ..., $H_{\omega 2} = 4n$, ..., $H_{\omega 3} = 8n$. Only by the time it reaches H_{ω^3} does it arrive at the level of tetration. But the Hardy hierarchy is historically important because it was the first time transfinite ordinals had been used to index a recursive function and so is the ancestor of much more rapidly growing families of functions such as the fast-growing hierarchy we described above. In fact, the two hierarchies are closely related. For any ordinal α, f_α is the same as H_{ω^α}, so that, for instance, $H_{\omega^{(\omega+1)}}(64)$ is comparable to Graham's number. The Hardy hierarchy eventually catches up with our fast-growing hierarchy – but not until its index reaches ε_0.

Googologists, whose pastime is the quest to define ever-larger numbers, prize fast-growing hierarchies because they're also used, and were first developed more than a century ago, in professional mathematics. Having been tried and tested in academia for many decades, they provide a solid foundation upon which enthusiasts can build more speculative ideas. In mainstream maths, fast-growing hierarchies are known, of course, to generate big numbers very quickly. But their main purpose is to serve as a benchmark for measuring the growth rate of other functions. Googologists, too, use them in this way, as a well-established and respected tool for gauging the strength of different types of functions that also mushroom with incredible speed. Notable among these explosively growing rivals is something called the TREE function.

As the name suggests, a tree in mathematics can have the appearance of a tree that grows in the ground or a family tree, with branches spreading out from a common trunk. Mathematical trees are a special variety of what in maths are known as graphs. Usually we think of graphs as being charts drawn on graph paper, in which one value is plotted against another. But the types of graphs we're talking about in connection with trees are different: they're ways of representing data in which dots, called nodes, are connected by line segments, called edges. If it's possible to start at a node, move to other nodes along edges, and then return to the starting node without repeating any edges or nodes, then the route taken is known as a cycle and the graph is said to be cyclic. If it's possible to start at any node and travel to another node, without repeating an edge or node along the way, then the route taken is called a path and the graph is said to be connected. A tree is defined as a graph that is connected but has no cycles. Both family trees and biological trees also have this kind of structure. If a unique number or colour is assigned to each node, then the tree is said to be labelled. Furthermore, if we assign one node to be the root, then we have a rooted tree. One useful property of rooted trees is that for any node, we can always trace back a path to the root.

Some mathematical trees that have the same kind of branching structure as a real tree can be fitted inside others of their kind. They're said to be homeomorphically embeddable, which is a fancy way of saying they're similar in form or appearance and one of them is like a smaller version of the other. Of course, mathematicians are a bit more precise about the definition. They start with a larger tree and see how much of it can be pruned using a couple of different methods. First, if there's a node (except for the root node) that has just two edges, either leading into or away from it, that node can be removed and the two edges joined into one. Second, if two nodes are joined by a single edge, the edge can be collapsed and the two nodes compressed into one. The colour of this new node is the colour of whichever node was originally closer to the root. If a smaller tree can be made by applying these two steps in any order to the larger tree, the smaller one is said to be homeomorphically embeddable in the larger. In 1960, the American mathematician and statistician Joseph Kruskal proved an important theorem to do with this kind of tree. Suppose there's a sequence of them so that the first tree can have only one node, the second up to two nodes, the third up to three, and so on, and that no tree is homeomorphically embeddable in any subsequent tree. What Kruskal found is that such a sequence always has to end at some point. The question is: how long can the sequence be? In response, American mathematician and logician Harvey Friedman, listed in *The Guinness Book of Records* in 1967 as the world's youngest professor (an assistant professor at Stanford aged just eighteen), defined the tree function, TREE(n). Friedman then investigated the output of the function for different values of n. The first tree consists of a single node of a certain colour, which can't be used again. If $n = 1$, this is the only colour and the sequence immediately stops, so that TREE(1) = 1. If $n = 2$, there's one more colour. The second tree can contain up to two nodes, so it contains two nodes both of this colour. The third tree also must contain only this colour, but can only have one node, as otherwise the second tree would be homeomorphically embeddable in the third. Beyond that, no more trees are possible so TREE(2) = 3. The big shock, as Friedman found, comes when we reach TREE(3). In a sudden explosion of complexity and proliferation the number of nodes far surpasses Graham's number and reaches the small Veblen ordinal, that extraordinarily un-small number we encountered in our travels among the various infinities, in the fast-growing hierarchy.

Humongous though TREE(3) is, it's definitely finite – as are all the numbers generated by the TREE function for any specific n. Kruskal himself showed that any TREE(n) will ultimately result in a tree that contains a previous tree, so that for every n, TREE(n) will output a finite result. Proving this for any particular TREE, however, is neither easy nor quick. Friedman found a way to calculate how many symbols, such as plus or minus signs, exponents, or other mathematical symbols, it would take to prove that TREE(3) was finite. His answer: $2 \uparrow\uparrow 1{,}000$, or a power tower of 2s stacked a thousand high.

The rate at which TREE(n) grows is tremendous, bounded below by the fast-growing hierarchy of the small Veblen ordinal. Even the TREE function, however, is totally outstripped in its growth rate by other functions that arise from graph theory. Friedman, who defined TREE, has also devised what are known as the subcubic graphic (SCG) and the simple subcubic graph (SSCG) functions. Like TREE, these start off modestly enough but then suddenly and sensationally inflate. For example, SSCG(0) = 2 and SSCG(1) = 6. Then there's a rapid growth to SSCG(2), which is approximately $10^{3.5775 \times 10^{28}}$ (and ends with the digits 11352349133049430008). Thereafter, all contact with numbers for which there's any conventional form of notation is lost. The exact value of SSCG(3) is unknown but it's been shown to be very much greater than TREE(3) or, for that matter, TREE$^{\text{TREE}(3)}$(3).

Despite the fact that TREE, SSCG, and super-fast-growing functions like them produce astonishingly large numbers very quickly, they are nevertheless *computable*. In other words, it's possible to write down a finite, step-by-step list of instructions, or algorithm, to calculate their output for any given input. However, there are other functions that can't be evaluated by a predetermined method, no matter what resources, such as time, memory, or processing speed, are available. They're said to be *uncomputable* and among them are functions that eventually outpace and dominate every one of those we've looked at so far.

CHAPTER 11

Does Not Compute!

BEAVERS ARE KNOWN for their industriousness and for building the largest structures in the animal kingdom. A family of four beavers can put up 1.5 metres of dam wall in a day and end up with a structure that's typically 5 to 10 metres wide and about 1.8 metres high. The largest beaver dam in the world, discovered in a remote part of northern Alberta, is an incredible 850 metres long and can be seen from space. With such edifices and the hard-working rodents that build them in mind, Hungarian mathematician Tibor Radó named one of the fastest-growing functions ever conceived – the 'busy beaver'.

To understand what the busy beaver function is all about, and why it's so impressive, we must delve a bit more into the subject of computability theory. In particular we need to take a closer look at Turing machines and the difference between what's theoretically computable and what isn't.

Alan Turing was among a group of mathematicians and logicians, including Alonzo Church, Kurt Gödel, Stephen Kleene, Rózsa Péter (whom we met earlier in connection with a simplified form of the Ackermann function), and Emil

Post, who pioneered computability theory in the 1930s. At the heart of computability theory is the study of computable functions – functions that take natural numbers as their input and produce an output that is possible (given sufficient resources) to calculate. In Chapter 8 we talked about the device that Turing invented as a model for computation. A Turing machine, remember, uses an infinitely long tape, divided into squares, which usually start off blank but onto which can be written a 0 or a 1. At a given moment in time, a read-write head is positioned over one of the squares and the machine can perform any of three basic actions: read the symbol on the square under the head; edit the symbol by writing a new symbol or erasing it; or move the tape one square to the left or the right. What it actually does depends on the contents of the present square and the instruction given in the machine's program.

Let's say that we have a Turing machine that has just two possible states, A and B. In state A it will always write a 1 to the tape, move the tape to the right, and change to state B. In state B, it will always write a 0 to the tape, move the tape to the right, and change to state A. It's pretty clear what will happen: the machine will never halt and will keep writing the sequence 101010… forever. Most Turing machines are more interesting than this because their instructions tell them first to read what value is on the current square and then change their behaviour accordingly.

In May 1962, Tibor Radó wrote an article called 'On Non-computable Functions' for the *Bell Systems Technical Journal*. In it he introduced his busy beaver function in the form of a game. The object of the game is simple: to find the maximum number of ones a Turing machine with n states can write on its tape before halting. It's easy to design a machine that writes 1s without ever stopping. All you have to do is give it a single state A with the instruction to write a 1 in the current square (whatever may be there already), move the tape to the left, and remain in state A, so that it will repeat the process endlessly. Radó's game, though, insists that the machine must eventually halt. It's also easy to come up with a machine that will write as many 1s as you like. If you want to write three 1s, you could have a machine with three states, A, B, and C. Each state tells the machine to write a 1 and then move the tape one square to the left. As well as this, state A will change to state B, state B will change to state C, and state C will halt. By using the same strategy but with more states, you could make a machine that would write 10, 50, a million, or as many 1s as you wanted on the tape. Radó's busy beaver game, though, is more subtle than that. It asks for the maximum number of 1s that can be written by a Turing machine with a *specific number of states* before halting. If $n = 3$, for instance, this maximum number is the third busy beaver number.

For very small values of n, the busy beaver numbers are also small and aren't hard to calculate. A Turing machine with only one state can write at most a single 1 before halting; the only alternative is for it to write 1s forever, either to the right or left, which violates the rules of the game. Surprisingly, there are 64 possible one-state Turing machines when all the different combinations of moving left or right, halting or not, or overwriting or leaving the same are considered. But the maximum number of 1s any of them can write *before halting* is one. So, the first busy beaver number is 1.

The greatest number of 1s a machine with two states can write, it turns out, is four, which can be achieved in as little as six steps. We can say that the winner of the two-state busy beaver game – the two-state busy beaver – has a winning output of four 1s so that the second busy beaver number is 4. You might suppose that this would be reasonably easy to figure out. But there are 20,736 possible two-state Turing machines, each one of which has to be checked before a winner can be declared. As Radó showed in his 1962 paper, the number of n-state Turing machines is given by the formula $(4(n + 1))^{2n}$. Putting $n = 3$, in order to figure out the number of three-state machines, the answer turns out to be 16,777,216. Every one of these has to be looked at to see if it might be a winner of Radó's game – in this case, to find the third busy beaver number. The machines that, at a glance, obviously won't halt can be discounted immediately, but that still leaves millions of others, which have to be checked. It's doable with the help of a computer, plus some clever hand-sifting, and the third busy beaver number turns out to be… 6. It seems like a lot of work to reach a very small number. With four states, the number of possible machines is about 25.6 billion, where again every one of which has to be scrutinised in one way or another to find out, first if it halts, and second, if it does, how many 1s it produces by the time it halts. The maximum number of 1s a four-state Turing machine can write, it transpires, is just 13.

So far, the busy beaver numbers are hardly impressive in size. The first four of them are 1, 4, 6, and 13, respectively. After all the effort of, in the case of the four-state machine, going through billions of possibilities, the biggest (finite) output is still only in the teens. But we've seen this kind of thing before, where a fast-growing function starts off at a pedestrian pace but then suddenly zooms away at a dizzying rate. We saw it with, for instance, the fast-growing hierarchies and the TREE function. It's also exactly what happens with the busy beaver.

The problem in progressing further is the explosive growth in the number of possible Turing machines (or Turing machine programs) as n gets bigger. The number of ways to program a five-state machine can be calculated from Radó's formula, $(4(n + 1))^{2n}$, with $n = 5$. It comes out to be $24^{10} = 6.34 \times 10^{13}$, or just over 63 trillion. That's a tremendous number of different options to have to check through, even if you have a supercomputer available (which some busy beaver researchers do). The problem isn't just the large number of combinations of states that a five-state machine can occupy but the fact that, in some cases, a single run may take a very long time before it halts, assuming that it ever does. Its tape may go this way and that, while the head writes long or short stretches of 1s only then to overwrite them, before beginning some new pattern of behaviour. Even after millions or billions of steps it may be impossible to tell if it will eventually stop or carry on indefinitely.

In the August 1984 edition of *Scientific American*, the monthly 'Computer Recreations' column by Alexander K. Dewdney was given over to the search for busy beaver numbers. Amateur mathematician George Uhing, from the Bronx, New York, was so inspired by reading it that he built a special-purpose piece of hardware, costing less than $100, to simulate a five-state Turing machine. After letting his homemade computer run for about three weeks while it sifted through several million possibilities, Uhing found one five-state machine that had printed 1,915 1s when it halted having gone through more than two million moves. His discovery set a new lower limit for the fifth busy beaver number, the previous best having been 501 1s.

Mathematician Allen Brady, of the University of Nevada at Reno, who verified Uhing's result, remarked on the significance of it being such a big jump from the four-state number (13). 'Our ability', he said, 'to distinguish between the machines that halt and machines that don't has diminished.' Dewdney remarked on the nature of the discovery: 'To me, the interesting thing is that basically you've got an amateur doing something which is of great interest to professionals.'

In 1989, German mathematicians Heiner Marxen and Jürgen Buntrock raised the lower limit from Uhing's 1,915 to 4,098, a record that still stands. In more than three subsequent decades, no one has found a five-state machine that halts after producing more than 4,098 ones, so this may well be the actual fifth busy beaver number. However, at least ten machines, which have been running for a long time, are still being checked. The likelihood is that these hold-outs will never halt, but the jury is still out.

From here on, the problem of computing busy beaver numbers becomes more or less hopeless. There are 28^{12}, or about 2.32×10^{17} (232 thousand trillion), six-state Turing machines, and with every increase in the number of states comes a bewildering rise in how many steps some of the machines may take before halting (if they ever do). No one knows what the sixth busy beaver number is and it'll be a long time before anyone ever does. The current six-state champion was found by Pavel Kropitz in 2010 and produces about $3.52 \times 10^{18,267}$ 1s after more than $7.41 \times 10^{36,534}$ steps.

Radó called his busy beaver function – the function whose output is the busy beaver numbers – Σ (sigma). So far we know that $\Sigma(1) = 1$, $\Sigma(2) = 4$, $\Sigma(3) = 6$, $\Sigma(4) = 13$, $\Sigma(5) \geq 4{,}098$, and $\Sigma(6) > 3.52 \times 10^{18{,}267}$. Based on the same reasoning that Kropitz used to set a lower bound for $\Sigma(6)$, the known lower bound for $\Sigma(7)$ is around $10^{\wedge}10^{\wedge}10^{\wedge}10^{\wedge}18{,}705{,}353$, and this will almost certainly increase with time. If you fancy your chances of getting into the record books by proving any of the busy beaver numbers beyond $\Sigma(4)$, or just finding a new lower bound, there's a thriving online busy beaver community, which anyone can join.

Taking a glance at what we know about the first seven busy beaver numbers, it's obvious that, after a slow start, they hurtle away into seriously big-number territory very quickly. Other fast-growing functions, like that of Ackermann, also tend to do this – lulling us into thinking that not much is happening before suddenly zipping off at terrifying speed into the numerical stratosphere. But the busy beaver function is different from any of the other functions that we've looked at so far. It's not only fast-growing and, after a few steps, difficult to compute by any practical means, it's actually *uncomputable*.

You may say, well, we've already computed the first four busy beaver numbers and there's a good chance we know, or will know before long, what the fifth one is. With faster and faster computers, and cleverer techniques, we'll progress further, and learn what the sixth busy beaver number is, and then the seventh, and so on. All that may be true – up to the 'and so on'. By brute force and smart thinking we can power our way up to some point on the busy beaver sequence: maybe seven, maybe eight. But that's not the point. The function, in general, is fundamentally uncomputable.

Remember our definition of a computable function: it's one that can be evaluated using an algorithm – a precise set of instructions, laid down in advance, that provides a sure-fire way of finding the value of the function for any given input. Think about a power tower of a hundred 10s: 10 to the 10 to the 10 to the 10, etc., a hundred times over. It's a colossal number, far bigger than a googolplex, and much too big ever to write out in full. But it's easy to write an algorithm that contains step-by-step instructions on how to compute it. The same is true of Graham's number, which is vastly larger. Just because we can't *actually* compute the value of a function for a particular input doesn't mean it's not computable!

However, no algorithm is available for computing busy beaver numbers. This doesn't mean that, at least for small inputs, we can't find some other method of calculating the value of the busy beaver function – such as trial and error. But if it's impossible to write down a finite sequence of well-defined instructions that can be carried out to compute the value of the function for *any* given input, the function is uncomputable.

Notice that this isn't a question of available resources. Faster computers and smarter software applied over long periods of time might allow you to use brute force – checking all the possibilities, one by one – to prove the next busy beaver number. But, as Radó demonstrated in his 1962 paper, you'll never find an algorithm to do the job for any arbitrary input, because it can be shown by logical argument that none exists.

One of the remarkable consequences of the busy beaver numbers being uncomputable is that they grow faster than *any* computable sequence of numbers. For short distances, early on, you can certainly find computable sequences that are in the lead. For example, the first few cube numbers are $1^3 = 1, 2^3 = 8, 3^3 = 27, 4^3 = 64$, all of which match or exceed the corresponding busy beaver numbers. Power towers of 10, say, can maintain a lead over the busy beavers a bit longer, before being overhauled. Even fast-growing, *non-primitive* recursive functions, like the Ackermann function, which are nevertheless computable, are eventually completely over-whelmed by the busy beaver function.

Although we don't know the value of any busy beaver numbers beyond the first handful, mathematicians are inter-ested in them because they can be used to gauge the difficulty of resolving some important, long-standing open questions. One of these is Goldbach's conjecture, which asks whether every even integer greater than 2 is the sum of two primes. In 2015, an anonymous user of the GitHub software develop-ment site published a program for a 27-rule Turing machine that would halt if, and only if, Goldbach's conjecture were false. This means that if BB(27) could be computed it would provide an upper limit on how long it would take to settle the conjecture automatically. All we'd have to do is run a 27-rule Turing machine up to a maximum of BB(27) number steps and if it hadn't halted by then, we'd know that Goldbach's conjecture was true. We could never do this in practice because BB(27) is so incomprehensibly huge. Nevertheless, being able to use busy beaver numbers in this way helps calibrate the state of our knowledge of unsolved problems in number theory. A similar analysis, carried out in 2016 by Scott Aaronson, in collaboration with Yuri Matiyasevich and Stefan O'Rear, identified a 744-rule Turing machine that halts if and only if the mighty Riemann hypothesis is false.

So far we've considered only Turing machines that operate with two symbols, 0 and 1. The busy beaver sequences that result from machines that work with more states grow at an even faster rate. If n specifies the number of states and m the number of symbols, then $\Sigma(n,m)$ is the busy beaver number for an n-state machine with m symbols. We already know that $\Sigma(2,2)$, or what we referred to previously as just $\Sigma(2)$, is 4. The table below summarises what's currently known about other busy beaver numbers. Focusing on the first column, look at how rapidly the size of the numbers increases, especially for $m > 4$, and what we are seeing in the table are merely the very beginnings of the growth. Imagine how stupendous must be a number such as $\Sigma(10,10)$ or the busy beaver number of a Turing machine with a googol number of states and a googol symbols, $\Sigma(10^{100}, 10^{100})$.

	Values of $\Sigma(n, m)$					
m ＼ n	2-state	3-state	4-state	5-state	6-state	7-state
2-symbol	4	6	13	4098?	$> 3.5 \times 10^{18267}$	$= 10^{10^{10^{10^{18705353}}}}$
3-symbol	9	≥ 374676383	$\geq 1.3 \times 10^{7036}$	?	?	?
4-symbol	≥ 2050	$> 3.7 \times 10^{6518}$	7	?	?	?
5-symbol	$\geq 1.7 \times 10^{352}$	?	?	?	?	?
6-symbol	$\geq 1.9 \times 10^{4933}$	?	?	?	?	?

It might seem that with busy beaver numbers we've surely reached the end of the road in terms of how fast a function can grow and how quickly it can reach into the deepest spaces of the numerical cosmos. Surely, with Radó's $\Sigma(m,n)$ we've arrived at the limit of what can reasonably be defined in our quest for the largest numbers of all. But no. The busy beaver function may be uncomputable but it's not unsurpassable. Time now, to enter the odd, outrageous, and occasionally obsessive realm of extreme googology.

CHAPTER 12

The Strange World of the Googologist

SOME PEOPLE LIKE to race fast cars, others to skydive or climb sheer rock faces. Googologists get their kicks from dreaming up ways to express larger and larger numbers. In some ways it's the ultimate geeky pastime. It's easy to imagine a young Sheldon Cooper, from *The Big Bang* series, spending hours engrossed in googology, competing online with other similarly inclined whizz-kids to name and define numbers of ever more outrageous size.

Googology has venerable roots, stretching all the way back to Archimedes and his big-number musings in *The Sand Reckoner*. The googol itself was named a century ago and the topic of large numbers became a familiar one in popular-level maths books beginning in the 1940s. Googology falls under the umbrella of recreational maths and is pursued for fun rather than with the expectation that it'll turn up anything of great importance. Research mathematicians, in their professional capacity, have little interest in numbers simply because of how big they may be. Their focus is on finding new proofs and principles, and solving problems, both pure and applied. Only in some special areas of the subject, such as computability and the theory of fast-growing functions, as we've seen, is there some overlap with the quest that galvanises the large number enthusiast.

Of course, googologists are also often, during the day, involved with mainstream maths as well. The truth is, you simply don't do this kind of thing in your spare time unless mathematics turns you on! We've already come across some mathematical heavyweights who've made contributions to the study of large numbers, including John Conway, Donald Knuth and Hugo Steinhaus. In some cases, this has happened as an unexpected spin-off of the main aim of their research. Ronald Graham, for instance, didn't start his work on solving an obscure problem in Ramsey theory with the intent of finding the immense number that's now named after him. Wilhelm Ackermann's sole aim in devising his eponymous function was to break new ground in computability theory by identifying a function that wasn't primitive recursive. It just turned out that the Ackermann function does lead to incredibly large numbers very quickly, though this fact isn't even mentioned in the original paper.

For the most part, googology is a sideshow, an intellectual hobby, pursued by people who are often students or graduates in maths or computer science and who have caught the big-number bug. The more prominent googologists run blogs or contribute frequently to websites and forums that are devoted to the pursuit of gigantic numbers. Among them is Adam Goucher, now a researcher in the Statistical Laboratory at Cambridge University. As a PhD student at Trinity College, Cambridge, Goucher mentored one of the authors (Agnijo) in his preparations to take part in the International Mathematical Olympiad (IMO). Goucher was a member of the UK team at the 2011 IMO, and Agnijo earned a perfect score of 42 points out of 42, along with a gold medal, at the 2018 IMO, before starting his maths degree in the same year, also at Trinity.

Goucher invented something he calls the Xi (Ξ) function. This arises from an area of maths known as combinatory logic, which was introduced in 1924 by Russian logician Moses Schönfinkel, and further developed by American logician Haskell Curry a few years later. The original purpose of combinatory logic was to get around the need for quantified variables in mathematical logic and thereby to understand better some paradoxes that had recently arisen out of set theory. It's closely related to the lambda calculus of Alonzo Church and, like it, has been used in computer science as a model for computation.

Goucher based his Xi function on *combinators*, which are the fundamental elements of combinatory logic and are the result of functions acting on themselves. The Xi function's claim to fame is that it grows faster than Radó's busy beaver numbers. In fact, Xi is one of the fastest-growing functions ever defined – though it can be, and has been, surpassed by functions that outpace it.

Another well-known googologist is the American amateur mathematician and self-described 'mega-arithmologist' who goes by the pseudonym Sbiis Saibian, a name he generated by randomly hitting letters on a keyboard. In 2008, at the age of twenty-five, while still a college student, Saibian launched a web book called 'One to Infinity: A Guide to the Finite', which has since become a popular resource in the community. Saibian is among the most prolific coiners of googolisms – unofficial and frequently humorous names for specific large numbers. He defines googology as 'the practice/ craft of generating, comparing, and naming large numbers and the study and theory of how to do so'.

Fellow American amateur mathematician (though with a masters degree in the subject from the University of Texas), famed for his large-number work, is Jonathan Bowers. Now in his fifties, Bowers is something of an elder statesman of googology and a pioneer of it as a subject with its own online following and self-invented rules and nomenclature. Bowers was inspired to delve into his own large-number investigations, he said, by reading a book on hyperoperators. All of amateur googology starts, in this way, from a secure basis in conventional maths – the hyperoperator series, fast-growing hierarchies, and so forth. Where it goes from there depends on the ingenuity and mathematical competence of the googologist. Some amateur googological work is no more than an extension of mainstream theory, while other results are more controversial and provoke debate and disagreements among the community of big-number celebrities and fans.

A certain protocol has emerged among googologists for what is acceptable and what isn't in putting forward ideas. Possibly the worst thing you can do in posting on a googology forum, apart from being abusive or otherwise obnoxious, is to propose what's become known as a 'salad number'. This starts from some existing number or function and consists of an uninteresting or inelegant extension of it without contributing anything new. The most obvious example is to say 'plus 1' after whatever large-number proposal someone else has made. So if we have a big-number naming contest and you start with 'Graham's number', I might counter with 'Graham's number plus 1', or 'Graham's number to the power of Graham's number', or even '$G_{64}\uparrow\uparrow\uparrow\uparrow\ldots\uparrow G_{64}$ with G_{64} up-arrows' (which is roughly G_{65}). But none of my replies are particularly clever or original.

It takes no great mathematical skill or leap of the imagination to jump to a slightly bigger number once any given number, however large, has been identified and clearly defined. There's always the next number down the number line, or one that's twice as far along. But if you start with Graham's number and I then apply the same kind of methods used in reaching Graham's number to hop to something bigger, I haven't really broken new ground: I'm still in Graham's number land. Any inelegant mishmash of existing numbers and functions put forward in an attempt to outdo some original large number or fast-growing function is a salad number and will attract scorn and derision from experienced practitioners of the googological art.

Having said this, a page has been set aside on one of the most popular large-number websites, Googology wiki (www.googology.wikia.org), solely for the enjoyment of those who want to indulge in mixing the most horribly complex and outrageous salad number on the planet: the mighty, useless, and wholly incomprehensible 'croutonillion'. The arbitrary starting point for this mathematical monstrosity is the already insanely large 1,234,567th busy beaver number. Given that we're still not sure what the fifth busy beaver number, $\Sigma(5)$, is (though it's thought likely to be 4,081), and all we know about the sixth is that it's at least $3.5 \times 10^{18,267}$, $\Sigma(1,234,567)$ is clearly indescribably huge. But the important point is that it's a well-defined specific finite number. On the croutonillion page, users are invited to add their own ingredient to the salad by suggesting a way of making the final outcome as big as possible. 'X' is used to represent the running total – in other words, the output that has resulted from all the previous steps. Starting from the 1,234,567th busy beaver number, the first operations that appear on the page are:

1 $X \uparrow X^X \uparrow X$

2 $\Sigma(X)$

3 megafuga(booga(X))

What on Earth do they mean? Step 1 involves substituting the 1,234,567th busy beaver number for X to give $\Sigma(1,234,567) \uparrow$ $\Sigma(1,234,567)$ $^{\Sigma(1,234,567)} \uparrow \Sigma(1,234,567)$. So, this is a straightforward hyperoperation represented in the form of up-arrows, albeit sensationally many of them. Remember in Chapter 4, when we were talking about how Knuth's up-arrow notation works, we saw what a surprisingly vast number $2 \uparrow\uparrow\uparrow 3$ was turning out to be as we started to unpack the arrows. After just a few lines of unpacking we were already confronted with $2\uparrow(2\uparrow(2\uparrow(2\uparrow(2\uparrow(2\uparrow(2\uparrow(2\uparrow(2\uparrow(2\uparrow(2\uparrow(2\uparrow65,536))))))))))))$, and the knowledge that the $2 \uparrow 65,536, = 2^{65,536}$, alone has around 7,400 digits when written out in full. $2 \uparrow\uparrow\uparrow 3$ is a mere pentation of two very small numbers but unpacks to something unfathomably large. The first step in making the croutonillion is to apply $\Sigma(1,234,567)$ $^{\Sigma(1,234,567)}$ up-arrows to the base $\Sigma(1,234,567)$ and power $\Sigma(1,234,567)$. The manically large output from this becomes our new value of X, which then feeds into step 2. All we need to do for this is to find the (new) Xth busy beaver number, or in other words, $\Sigma(\Sigma(1,234,567) \uparrow \Sigma(1,234,567)$ $^{\Sigma(1,234,567)} \uparrow \Sigma(1,234,567))$, which then becomes the X value to be put into step 3.

Googologists love nothing more than dreaming up silly names for their contrived super-fast-growing functions. Step 3, megafuga(booga(X)), involves the combined action of two functions, extrapolated from the conventional fast-growing hierarchy by Sbiis Saibian and defined and named by him on his Large Number site. So the merry madness continues, on and on down the eccentric list of croutonillion instructions. At step 12, we come across 'gongulus-(2X + 1)-plex', whatever that is. At step 13 is our friend the TREE function in the form TREE^XX. If you have the time and patience you can visit the croutonillion page and check out all 3,978 steps that have been entered so far. A good place to start, though, is with the very first sentence on the page: 'Croutonillion is a groundbreakingly pointless googologism, consisting of a ridiculous sequence of totally arbitrary steps.' You have been warned!

Mathematics is one of the most intellectual of pursuits. Certainly it has important practical applications and, indeed, its utility in such areas as commerce and architecture is what helped spur its rapid development among early civilisations. But *pure* mathematics – maths for maths sake – is an extreme challenge for the mind that's taken up irrespective of whether it might eventually be put to some 'real-world' use. Not surprisingly, given that mathematicians are among the most elite of abstract thinkers, there's a long history of them pitting their wits against each other in competitions. Such contests go back at least as far as the 1500s when in Renaissance Italy, rival mathematicians would pose each other problems and try to solve them, while others would bet on the outcome. Today's only official maths competitions are for young people, culminating in the annual International

Mathematical Olympiad. But googologists vie with each other to describe and name ever-larger numbers. Even in the academic world, challenges are sometimes issued in the form of puzzles to be solved and games to be won. Remember Pierre de Fermat's handwritten note in the margin of a book, discovered after his death, which was almost certainly a dare to other mathematicians to try to solve what became known as Fermat's Last Theorem.

In the last chapter, we saw how Tibor Radó, in a 1962 paper, introduced his busy beaver function in the form of a game. The object is to find the maximum number of 1s that an n-state Turing machine can write before halting. In fact, Radó's game, as well as exploring the behaviour of a particular uncomputable function, does have some significant connections with orthodox mathematics. Its importance in this respect has been highlighted by American theoretical computer scientist Scott Aaronson, who serves as David J. Bruton Centennial Professor of Computer Science at the University of Texas at Austin and as the founding director of UT Austin's Quantum Information Center. Aaronson is an example of a leading academic whose work spills over into googology. His essay 'Who Can Name the Bigger Number?' is widely referenced in academic computer science as well as being a good, easy-to-follow introduction to large numbers in general.

A more recent paper, titled 'The Busy Beaver Frontier', published by Aaronson in 2020, brings the story of Radó's function up to date and extols its virtue as a teaching aid:

> *In my opinion, the BB function makes the concepts*
> *of computability and uncomputability more vivid*

than anything else ever invented. When I recently taught my 7-year-old daughter Lily about computability, I did so almost entirely through the lens of the ancient quest to name the biggest numbers one can. Crucially, that childlike goal, if pursued doggedly enough, inevitably leads to something like the BB function, and hence to abstract reasoning about the space of possible computer programs, the halting problem, and so on.

Aaronson also explains how further busy beaver research may provide crucial insights into a number of open (unsolved) problems in number theory, set theory, and the foundations of mathematics. Among these is the Collatz conjecture – a hypothesis so simple to state that it can be used as a maths exercise for young children. It says that if you start with any whole number, divide it by 2 if it's even and triple it and add 1 if it's odd, and keep repeating these actions over and over again, you'll eventually end up with the number 1. Despite repeated efforts by number theorists since German mathematician Lothar Collatz first put forward the suggestion in 1937, it remains an open problem. French mathematician Pascal Michel and Aaronson have both argued that understanding the behaviour of Turing machines in busy beaver games, especially determining when they halt, is closely related to resolving the Collatz problem. Aaronson also points out that the sequence of busy beaver numbers tests the limit of the system of axioms – so-called Zermelo–Fraenkel set theory – that forms the basis of almost all modern mathematics.

So, although most professional mathematicians don't share the hobbyists' fascination with defining and naming ever more gigantic numbers for the sake of it, googology isn't without its merits. It may eventually help resolve some long-standing open problems, and it reveals the limits of our current mathematical universe, just as peering into space with the world's largest telescopes pushes back the frontiers of the physical cosmos.

It's hardly surprising that googology, which is all about going further and faster, has given rise to several contests. One of the first of these was the Bignum Bakeoff, organised by American mathematician David Moews, who took first place in the 1984 International Mathematical Olympiad with a perfect score of 42 points out of 42 and earned his PhD from the University of California at Berkeley in 1993. Competitors in the Bakeoff, held in December 2001, were challenged to produce the biggest number they possibly could from a computer program that was no more than 512 characters long (ignoring spaces) in the programming language C. No present-day computer could actually complete any of the programs submitted within the lifetime of the universe, so the entries were analysed by hand and ordered based on their position in the fast-growing hierarchy. The winning entry was a program called loader.c after its creator, the New Zealander Ralph Loader. A computer with an unfeasibly large memory and an outlandish length of time available would be needed to generate the final output. But if it could be done, the result would be Loader's number – an integer known to be bigger than TREE(3) and some other heroic inhabitants of the googologist's cosmos, such as SCG(13), the 13th member of the sequence of subcubic graph numbers.

In 2007, a large-number contest called the Big Number Duel pitted philosophers and old-graduate-school chums

Agustin Rayo (aka The Mexican Multiplier) of MIT and Adam Elga (aka Dr. Evil) of Princeton against each other in a back-and-forth tournament to see who could define the most colossal integer. The numerical slugfest, which blended comedy, convoluted mathematical, logical, and philosophical manoeuvrings, and the melodrama of a world-title boxing match, took place in a packed room in MIT's Stata Center. Elga opened optimistically with the number 1, perhaps hoping that Rayo would have an off day. But Rayo swiftly countered by filling the entire blackboard with 1s. Elga promptly rubbed out a line near the bottom of all but two 1s, turning them into factorial signs. Then the duel progressed, eventually transcending the bounds of familiar mathematics, until each competitor was inventing their own notations for ever-larger numbers. At one point, a spectator asked Elga 'Is this number even computable?' to which Elga,

LARGE NUMBER CHAMPIONSHIP
Two competitors. One chalkboard. Largest integer wins.

Sponsored by MIT Linguistics & Philosophy. For details see http://student.mit.edu/iap/nc19.html

Friday
Jan. 26
3pm
32-D461

Your MIT
DEFENDING CHAMPION
Agustín
"The Mexican multiplier"
"Plural power"
"Ray gun"
RAYO

The
CHALLENGER
Adam
"The mad Bayesian"
"Dr. Evil"
"Elg-finity"
ELGA

FIGURE 2.1: Promotional poster for the Big Number Duel.

after a brief pause, replied 'No'. Finally, Rayo delivered the knock-out blow with a number that he described as: 'The smallest positive integer bigger than any finite positive integer named by an expression in the language of first-order set theory with a googol symbols or less'. Just how large Rayo's number is we don't know and probably never will. No computer could ever calculate it, even given access to a universe that could hold a googol symbols or more.

Is Rayo's number the biggest number in the world? Clearly not, because there's always Rayo's number plus 1 – not to mention the messy numerical salad that is the croutonillion (and which includes many ingredients that build on Rayo's number). We also have to start thinking about the assumptions we make when we talk about very large numbers. Unlike the physical universe in which we live, there are many possible mathematical universes, each of which is uniquely defined by the set of axioms – the assumed truths – that determines its nature.

CHAPTER 13

Bridge to Beyond

THE NUMBER LINE stretches away forever. And yet, somehow, beyond it, lies something else – an infinite number of numbers that are themselves infinite in size. Between the two, there's no smooth region of transition: no numbers that are more than finite but less than infinite. This has been the reality of mainstream mathematics since the turn of the twentieth century, when Georg Cantor's work on set theory and transfinite numbers started to become widely accepted.

But there are those who object to this now-orthodox picture of mathematical reality. Infinity is a fiction, say some. There are limits to how large, in practice, finite numbers can be, argue others. Meanwhile, recent mathematical research has ventured into a no-man's land that may exist between finite numbers and infinity, bringing into question the meaning of infinity and the relationship between mathematics and the world in which we live.

Infinity has always been a touchy subject. By its very nature we can't imagine it. A brain that's finite in size can never form a meaningful mental image of the infinite. But that's hardly a good reason for denying it could exist. Most of us can't grasp what an object would look like in four or more dimensions (although a few claim to have trained themselves to do it), yet there's no question that theoreticians can quite happily deal with higher-dimensional shapes and spaces. Mathematicians regularly analyse all kinds of things that we can't picture in our mind's eye. For that matter, physicists have made discoveries about the real world that are unimaginable – for instance, most of what happens in the realm of quantum mechanics. But the limitations of our ability to visualise, or to understand through 'common sense', aren't good reasons to dismiss the possibility that something is real.

When it comes to infinity, we *can* picture something stretching away indefinitely into the distance – whether it be outer space or the number line. We can also appreciate the idea of time just going on and on, either in a linear or a cyclical way, and not abruptly stopping at some point. In fact, in many ways, when it comes to the physical universe, or the number line, it's just as hard to imagine them reaching an end as it is to picture what it would be like for them having no end. After all, if time were eventually to stop, what would come next? Existence seems to demand time. And if space ends at some boundary, what lies on the other side? If there were no such thing as infinity in maths, where and what are the limits beyond which we can't go?

Aristotle's distinction between potential infinity and actual, or 'completed', infinity, and his denial of the existence of the latter, has had an enduring appeal. Cantor was mercilessly attacked, most notably by his old mentor Leopold Kronecker, for daring to bring actual infinity into mathematics. Kronecker said: 'I don't know what predominates in

Cantor's theory – philosophy or theology, but I am sure that there is no mathematics there.' Well before Cantor appeared on the scene, Carl Gauss had made his position clear, along with that of most mathematicians at the time: 'Infinity is nothing more than a figure of speech which helps us talk about limits. The notion of a completed infinity doesn't belong in mathematics.'

Objections to the existence of infinity in maths rumbled on after set theory had become well established. Austrian-British philosopher Ludwig Wittgenstein, Dutch mathematician Luitzen Brouwer (founder of modern topology), and English mathematician Reuben Goodstein were among those who challenged the new status quo. The crux of this rival philosophy, known as finitism, is that the infinite is a fantasy conjured up by the human mind – an invalid concept in practice – and that only finite mathematical objects truly exist. Finitists don't dispute the reality of natural numbers and may be happy to accept that such numbers don't just suddenly hit a roadblock somewhere. They generally agree that the idea of the number line being unbounded is a reasonable idea and one that's essential, for instance, in the definition of a limit.

If we take as an example the graph of $y = 1/x$: as x gets bigger and bigger, y gets smaller and smaller, and the curve approaches the x-axis ever more closely. Conventionally, we say that the limit of y as x tends to infinity is zero. A finitist might not take issue with such a statement but would contend that introducing infinity in this way is just a shorthand or formalism, and that the only mathematical reality involved is finite numbers. In terms of set theory, finitism rejects the idea of the set of *all* natural numbers.

Philosopher and historian of mathematics Mary Tiles has drawn a distinction between what she calls classical finitists who, in the Aristotelian mould, are happy to go along with the notion of potentially infinite objects, and strict finitists. Someone in the latter camp would, for instance, have a problem with the statement 'every natural number has a successor' and the extent to which it's meaningful to talk about an infinite series being the limit of a partial finite sum. Taking finitism a step further leads to the stance known as ultrafinitism, which challenges even the suggestion that finite numbers can be indefinitely large.

At the heart of ultrafinitism is the belief that the physical universe is all that exists and that there's no Platonic place detached from the natural world inhabited by purely mathematical entities. An ultrafinitist may refuse to accept the existence of a number if there isn't, or never will be, enough space, time, matter, energy, or computational capacity to accommodate it. According to this philosophy, there's literally no room for many of the monstrously large integers we've encountered on our journey so far. Conventionally, a number such as $5 \uparrow\uparrow\uparrow\uparrow 8$ – a hexation expressed here in terms of up-arrows – is assumed to exist even though it would be impossible, physically, to write it out in full. An ultrafinitist would deny this assumption and point to the fact that the number could never be reached, in practice, by starting from 0 and applying the successor function $5 \uparrow\uparrow\uparrow\uparrow 8$ times. There's not enough space, matter, or time in the natural world to allow it to happen. Ultrafinitism, then, is a resource-based philosophy of maths, while finitism in general echoes Aristotle's view that while potential infinity is acceptable, actual infinity is not.

The young British mathematician Lawrence Hollom, who competed for the UK in the 2015 and 2016 International Mathematical Olympiads, has provided a somewhat tongue-in-cheek argument to define what might be the largest number that could possibly be instantiated in the physical universe. Hollom, who when not studying mainstream maths at the University of Cambridge is a googologist operating under the soubriquet Dr Ceasium [*sic*], has come up with something he calls the iota function, $I_m(n)$. The two variables of this function are the input (m) and the time (n).

Hollom defines n in his very physics-oriented and anthropocentric function to be the number of Planck times that have elapsed since the crossover from 1 BCE to 1 CE (there being no year 0). A Planck time, remember, is the smallest meaningful unit of time in nature, equal to about 10^{-44} seconds. Hollom's iota function stakes its claim to be able to produce the largest possible number by combining every function that has ever been defined, from 1 CE onwards, up to some specified point measured in Planck times (the ultimate salad number if you will). This may seem like a recipe for utter chaos since functions come in all different forms. However, Hollom has laid down four criteria, any one of which must be met before a function is eligible to be incorporated within the iota function. These criteria specify what to do with functions, depending on how they're initially set up, to render them into a form that lets them be included. If a function doesn't fit any of the criteria at the time to which I corresponds, it can still be factored in providing it's accessible in some type of physical storage, such as computer memory, a human brain, or writing.

Hollom's function is a broad church that even welcomes in all iotas from previous times. In other words, the set of functions comprising iota at time n, $I_m(n)$, contains all $I_x(n)$ for $x < m$. Duplicate functions are allowed too. The value of $I_m(n)$ reflects the combined action of all the functions at time m (in Planck units) since the start of year 1 CE, with n chosen to produce the highest possible output. Not all functions that are eligible have to be used if they fail to contribute to a maximum value.

Hollom's function at the present moment – the moment you're reading this – has a maximum value, even though we've no way of finding out what it is. The function is uncomputable, so that it can never actually be calculated. Nevertheless, it *does have* a definite value and there's no ambiguity about what this is. So now we're up to date, as it were. But we still have to take into account future iotas because this is where Hollom's *number* – the ultimate number in the universe (according to Hollom!) – comes from.

Future iotas, too, will have actual values, though they won't be determined until we reach the corresponding future times. This is because each future iota can include not only functions presently known, but also all functions that will arise between now and then, including all new iotas that spring into existence at every Planck time step. Hollom's function ensures that it will always be able to output the absolute upper limit of numbers that humanity can create, by absorbing any function or operation that surpasses it within the very next 10^{-44} of a second! Even if a similar rival to the iota function arises, the best it can do is maintain supremacy for one Planck time before the iota function subsumes it in the next cycle.

Hollom's number is the grand culmination of the action of the iota function evaluated over the entire possible history of human activity between 1 CE and a far future in which all the matter in the universe has collapsed into black holes, around 1.18×10^{54} years after the Big Bang. One suspects that the human race may not stick around that long: it's sometimes hard to predict 12 months ahead, let alone a million trillion trillion trillion trillion years. Nevertheless, Hollom's number is defined in this way: as the output of the iota function applied to the number 200 at a time of about 1.18×10^{54} years after the moment in which our universe came into existence, or whatever stabilisation point is reached prior to that. At present it's undetermined as it uses functions that haven't even been defined yet. But in the fullness of time it will take on a value that will never be exceeded.

Of course, it's a little parochial to suppose that humans are the only ones who'll invent new, more powerful functions in the future. Undoubtedly, if we survive, we'll build computers whose intelligence and mathematical prowess will far exceed our own. Then again there may be civilisations on other worlds, which make our ability to generate large numbers look like child's play. But it's easy to generalise Hollom's function so that it absorbs new functions, wherever and however they're devised.

Lawrence Hollom didn't intend for his iota function to be rigorously defined or to be taken too seriously. It's just an interesting thought experiment. In contrast, some recent research within professional mathematics sets up the possibility of a real breakthrough in understanding the finite–infinite divide. Mathematician Keita Yokoyama, at the

Japan Advanced Institute of Science and Technology, and computer scientist Ludovic Patey, at Paris Diderot University, have managed to shed light on a shadowy land between finite numbers and infinity. Their work also bears upon the relationship between mathematics and physical reality.

Broadly speaking, there are two kinds of statements you can make in mathematics: finitistic ones and infinitistic ones. As the names imply, finitistic statements can be proved without any recourse to the concept of infinity, whereas those that are infinitistic rest on the assumption that infinitely large objects exist. A central aspect of mathematical logic is to explore this division, which is why Yokoyama and Patey's work is significant. In 2016, the two young researchers came up with a proof, to do with a statement called Ramsey's theorem for pairs, that's finitistic in nature even though it applies to an infinite set of objects. Their result would have pleased David Hilbert, who, in 1921, tasked mathematicians with the goal of proving that all of mathematics could be grounded in a limited set of axioms, which could be proved to be consistent, or free from contradictions. An essential part of his Program was to supply finitistic proofs and justifications for all statements to do with infinity. Only by reducing everything to arguments involving finite maths, Hilbert believed, could the scepticism that swirled around Cantor's set theory and transfinite numbers be dispersed.

A century ago, there was still dissatisfaction among mathematicians stemming from the fact that presuming the existence of actual infinity did nothing to help with calculations. If abstract methods and concepts had no practical value – if they couldn't be used as part of a computation – then, critics argued, what was their reason for being? Even more of a threat to the new Cantorian regime was that it implied some outrageously non-intuitive ideas, such as that transfinite numbers come in an infinite hierarchy, apparently unconnected to physical reality, and a curious result announced in 1924 by Polish mathematicians Stefan Banach and Alfred Tarski. According to the Banach–Tarski paradox, you could take a ball, composed of infinitely many pieces, and reassemble those pieces to make two balls (or three or more) that are exactly the same size as the original. Of course, you couldn't do this mathemagical trick with an actual ball because matter in the real universe is made of atoms, which aren't infinitely divisible. But, in some ways, that strengthened the argument of critics: if the assumptions upon which a theory was based were wholly unrealistic, how could the theory be vindicated?

Concerns that the mathematics of the infinite were too abstract to be taken seriously or, worse, that they led to actual contradictions, spurred Hilbert to propose his Program. 'No one shall expel us from the paradise which Cantor has created for us,' Hilbert declared in a 1925 lecture. A sure way to avoid that eviction, he believed, was to put the theory of infinite sets on a firm logical footing by showing that it was reducible to finitistic proofs upon which everyone could agree.

In 1931, however, Hilbert's hopes were dashed. Austrian logician Kurt Gödel stunned the mathematical world by demonstrating beyond doubt that no system of axioms can ever prove its own consistency. An immediate outcome of Gödel's 'incompleteness theorems' is that the axioms of any system of finite mathematics can't even prove their own consistency, never mind the consistency of theories about the infinite.

Over time, this problem of the logical divide between finitistic and infinitistic maths has niggled away and, in some ways, grown more urgent. Based on the assumption that infinite sets exist, mathematicians have proved theorems about natural numbers that might be of relevance in physics and computer science. Most famously, in 1994, English mathematician Andrew Wiles drew on infinitistic logic in his proof of a statement that had confounded number theorists for more than three centuries – Fermat's Last Theorem.

The need to build bridges between the finite and the infinite hasn't gone away just because of the discovery that Hilbert's Program can never be completed. If his Program can be even partially realised then there may be many statements to do with infinite numbers of objects that it may be possible to connect, via finitistic reasoning, to the physical world around us. This is where the research carried out by Yokoyama and Patey on a result based on Ramsey theory is relevant.

Ramsey theory, in general, studies the conditions under which order must inevitably be present even in the most chaotic of systems. The various theorems that have sprung up under this large umbrella are based on the principle that within large, complex systems, however disorderly they may appear at first sight, there exist subsystems that have a definite structure. The implication is that there's no such thing as true randomness. We've already looked at an application of Ramsey theory in connection with Graham's number. The one at the heart of Yokoyama and Patey's research is Ramsey's theorem for pairs, or RT^2_2.

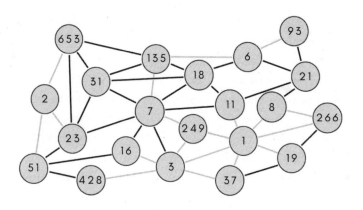

FIGURE 13.1: Ramsey's theorem for pairs of numbers (A, B), where $A < B$, using the rule: colour the connection black if $B < 2^A$, otherwise colour it light grey. If the set of pairs is infinite, Ramsey's theorem asserts that there will be infinitely many pairs with the same coloured connection.

One way to understand RT^2_2 is to imagine the set of all natural numbers, or at least a small part of it. Picture the positive whole numbers as being like little bubbles floating in infinite space. Each number is paired with every other, by a fine network of connecting lines as in Figure 13.1. Then imagine the line that joins any given pair of numbers as being coloured either blue or red according to some rule. For example, the rule might be: For any pair of numbers (A, B), where $A < B$, colour the pair blue if $B < 2^A$, otherwise colour it red. When all the colouring is complete, RT^2_2 says that there'll be an infinitely large subset of number pairs for which the connecting lines are the same colour. This result will hold true whatever is the rule chosen to determine the way the colouring is carried out. The breakthrough that Yokoyama and Patey made was to show that the proof of RT^2_2 can be rendered in terms of the logic of finite numbers.

Over the past few decades, logicians have found that the proofs of thousands of different theorems in mathematics can be reduced to just five major systems, or levels of logic. Levels one and two, the least powerful, are finitistic, whereas the rest are infinitistic. In 1972, it was shown that Ramsey's theorem for triplets, or RT^3_2, uses a proof from level three of this hierarchy, which depends on infinitistic methods. RT^3_2 involves colouring the links between *trios* of natural numbers (or other objects) in an infinite set either one colour or another, according to a predetermined rule. It turns out that the infinitely large single-coloured subset of triplets that RT^3_2 dictates is bound to exist after this has been done is too complex to be handled by reasoning based on finite sets. But what about its sister theorem, RT^2_2?

Some light on the status of RT^2_2 was shed in 1995 by British logician David Seetapun and American mathematician Theodore Slaman. They were able to show that RT^2_2 falls below RT^3_2 in the logic hierarchy because of the less intricate method of colouring involved. Further work, published in 2012, however, proved that neither did RT^2_2 belong to the logic system immediately below – the second level of the hierarchy – which is finitistic. The big question remained: where exactly did RT^2_2 fit in the logic scheme of things?

Yokoyama and Patey provided the answer. Using a combination of methods, they were able to show that the logical strength of RT^2_2 is equal to that of the arithmetic of primitive recursive functions – the types of functions that, as we saw in Chapter 7, we meet most commonly in maths. Crucially, this arithmetic is finitistically reducible. It's a fascinating and important result because it shows there are theorems that effectively span the finite–infinite divide. In logical strength, RT^2_2 falls somewhere between levels two (finitistic) and three (infinitistic) in the hierarchy of main logic systems. Yet although it's a statement about an infinitely large collection of things, it can be proved using logic that doesn't invoke infinity. In fact, Ramsey's theorem for pairs is reckoned to be the most complicated statement involving infinity that is known to be finitistically reducible.

Mathematicians may now be able to use the same approach, deploying the infinite apparatus of RT^2_2 to establish further proofs in finitistic maths, and thereby build more bridges between the finite and the infinite. Such developments would mark at least a partial realisation of Hilbert's Program. What's more, for those reluctant to acknowledge the reality of infinity in maths there's also some comfort in Yokoyama and Patey's discovery. RT^2_2 and its bi-colourable infinite sets may not translate directly to anything in the real world. However, the fact that its logical basis is on the same level as the maths that underpins most of science shows mathematical infinity can be allowed without necessarily requiring infinitistic logic or proofs.

The debate about whether finitistic reducibility strengthens the case for actual infinity rumbles on. It's probably true to say that most mathematicians are happy to go along with the utility of infinite sets and not get involved with deep philosophical discussions about whether or not infinity really exists, either in mathematics or the physical world. For our purposes, let's accept the standard view of transfinite numbers as developed by Cantor. It's time to answer the question with which we started: what is the biggest number in the world?

CHAPTER 14

The Biggest Number of All

WHERE DOES ALL this leave us? To the question 'What is the biggest number in the world?', what's the best answer we can give? As you'll have guessed by now, it's complicated! For one thing, there are both finite and transfinite numbers and it's hardly fair that they should be in the same competition. But even within the realm of the finite, when very large numbers are defined in completely different ways it can be hard to gauge their relative size. To add to the fun, there's the issue of what starting assumptions we make – in other words, the set of axioms that we take to form the basis of our arithmetic.

When we started our quest for the biggest number, matters seemed clear-cut. We're all familiar with the number line, and it's easy at the beginning to tell which of two numbers is the larger. Ten is bigger than one, 100 is bigger than 10, a googolplex is bigger than a googol, and so on. We quickly exhaust our ability to *imagine* big numbers, but that doesn't matter. No one can truly grasp in their mind's eye the immensity of a googolplex, but every child knows that a googolplex plus one is bigger than a googolplex.

Relatively few numbers have special names. Some of these, such as googol and moser, are invented, while others, like Skewes' number and Graham's number, honour their discoverer. They serve as useful markers as we venture further and further down the number line. But we quickly run into the problem of how to represent them. Exponentials work fine for the kind of ordinary big numbers we meet in everyday life or most of science. There are fewer than a googol subatomic particles in the observable universe and the googol is easily written as 10^{100}, where the 100 refers to the number of zeros if we were to write it out in full. The only numbers in physics or cosmology that put a strain on the familiar exponential form are those involving Poincaré recurrence times for the entire universe – the times estimated for all the contents of the universe to return precisely to some earlier quantum state. The largest of these ever to appear in print is 10^10^10^10^10^122.

So much for the physical world and numbers that are relevant to the circumstances of matter, energy, space, and time in which we find ourselves. Mathematics, on the other hand, isn't confined to dealing with questions that are relevant in nature. What's more, if you assume, as most mathematicians do, that there's a Platonic realm where mathematical objects exist outside of space and time, there's no limit to the room available for writing any number in any form we like. Take Graham's number, for instance. We know that it's a specific positive integer and that it ends with the digits 195387. We also know that it's 'just' a power tower of 3s – although a fantastically tall one. In the Platonic realm, free from any physical limitations, Graham's number is presumed to exist, complete, now and forever, expressed in full or as a power tower with all the 3s stacked unimaginably high, or in any other form we'd care to represent it.

The problem is, when we bring things over from this unbounded region of pure intellect and try to instantiate them in the real universe we run into constraints. Graham's number can't *actually* be written out in full. It can't *actually* be written as a power tower of threes either. There isn't nearly enough space or matter in the observable universe to accommodate it in these expanded forms. In any case, if we're looking at it from a practical and human perspective, writing out gigantic numbers, even ones much smaller than Graham's number, in full, would take up not only a lot of space and material, but enormous amounts of time. And what would be the point? No one would be able to make use of so much detail or mentally grasp it.

In the case of Graham's number, you may remember, we ended up calling it g_{64}, which is nice and compact. But special names, or abbreviations, can only be given to a tiny handful of all the numbers that exist because there aren't enough meaningful symbols (letters, numbers, and so forth) to go around. Also, whenever a shorthand representation is used, for it to be mathematically meaningful, it has to be backed up by a precise definition of what the number is and how (at least in theory) to calculate it.

We discovered, earlier in the book, that the mathematical operations with which we're familiar from our schooldays, including counting up by one (succession), addition, multiplication, and exponentiation, are just the beginning of an unending hierarchy of increasingly powerful mathematical actions known as the hyperoperator sequence. Beyond exponentiation lie tetration, pentation, hexation, and so on forever, each operation amounting to the repeated action of the one immediately below it in the hierarchy. It takes some ingenuity to find convenient ways of representing these increasingly powerful operators when we move beyond exponentiation. It's easy enough to write 3 to the power 4 as 3^4 and we know that, written out in full, this equals 81. But a more flexible scheme is needed if we want to represent, say, the hexation of 3 by 4. One such scheme is Knuth's up-arrows, in terms of which, for instance, 3 hexated to the 4 is shown as 3 ↑↑↑↑ 4. Despite its innocuous appearance, this number is far too big to be written out in full in the real universe. Yet, as we learned, even the compact notations available for writing the results of hyperoperator sums are hopelessly inadequate when faced with anything as large, or larger than, Graham's number.

Fortunately, help is at hand when up-arrows and similar systems of notation are overrun. The trick is to harness a higher level of recursion, or feedback, so that the number of times a repetitive action has to be carried out isn't fixed at the outset but becomes known only as the calculation unfolds. This leads to the concept of non-primitive recursive functions, such as Ackermann's function and the equivalent scheme introduced by John Conway – chained-arrow notation.

Bearing all this in mind, where do we stand in our journey towards the 'biggest number in the world'? Let's put in order of size, from smallest to largest, some of the named big numbers we've met in this book, starting with a million, which we can just about visualise in its entirety (imagine the crowd at a couple of Woodstocks) and can easily write out in full.

One million, 10^6 = 1,000,000

Avogadro's number, 6.022×10^{23}

Archimedes' sand number, ~10^{63}

Number of fundamental particles in the observable
universe, ~10^{86}

The googol, 10^{100}

One centillion, 10^{303}, the largest number ending in
-illion named in dictionaries

Largest known prime number, $2^{82,589,933} - 1$ (over 23
million digits long)

The googolplex, $10^{googol} = 10^{10\wedge100}$

Skewes' (first) number, $10^{\wedge 10 \wedge 10 \wedge 34}$

Poincaré recurrence time for the universe,
$10^{\wedge 10 \wedge 10 \wedge 10 \wedge 10 \wedge 1.1}$

Up to this point, the ranking is straightforward. All we have to do is look at the size of the exponential and organise the numbers so that the power of 10 increases from one to the next. If the powers are stacked, then, generally speaking, the higher the stack the bigger the number. Also notice that three of the numbers in the list are from the worlds of chemistry, physics, and cosmology, respectively, and two of these are estimates. After the cosmic Poincaré recurrence time – the largest value ever referred to in a scientific paper – all the numbers going forward will be purely mathematical with no known counterpart in science.

As we extend the list, we must also leave behind any familiar ways we learned about in school for writing numbers, including scientific notation. We must abandon the use of exponents – even stacked exponents or power towers – because the numbers are simply too large to be represented in this way. Instead we have to resort to arrow notations, or their equivalents. The next three milestones we reach on our trek to find the biggest number of all come from the work of Steinhaus and Moser and their system that involves writing integers inside polygons, nested polygons, and circles:

The mega, between 10 ↑↑ 257 and 10 ↑↑ 258
The megiston, between 10 ↑↑↑ 11 and 10 ↑↑↑ 12
The moser, between 2 ↑$^{mega-2}$ 3 and 2 ↑$^{mega-2}$ 4

For the mega and megiston we've switched to Knuth's up-arrow notation because we've moved into the realm of tetration and beyond. The mega is vastly bigger than anything we'd sensibly want to write in exponential form. Recall what a double up-arrow means: a power tower of the number before the two arrows, the height of which is given by the number after the arrow. So, 10 ↑↑ 257, which is smaller than the mega, is 10^10^···^10, where the three dots indicate another 255 levels of 10. Clearly, the mega is a heck of a lot bigger than Don Page's cosmic Poincaré recurrence time, which is only a bit more than 10 ↑↑ 4.

We can tell, at a glance, that the megiston dwarfs the mega because of the extra up-arrow. Pentating 10 to the 11 produces a number much, much bigger than tetrating 10 to the 258 (despite the fact that 258 is more than 11). If we were to write the megiston in terms of double up-arrows, it would be

10 ↑↑ 10 ↑↑ 10 ... 10 ↑↑ 10 ↑↑ 10

where there are eleven 10s in all, separated by double up-arrows. This obviously far exceeds $10 \uparrow\uparrow 257$ or $10 \uparrow\uparrow 258$, between which the mega lies.

But then we come to the moser, and an immense leap in size – far bigger than anything that's come before. To calculate the moser, we have to operate with a mega (minus 2) number of up-arrows. We've clearly entered into a realm of numerical giants that makes everything we've looked at before appear negligible and almost indistinguishable from zero. Moser is our current champion, though perhaps not for long because next it must come up against the mighty Graham's number.

In Chapter 5, we built up Graham's number in stages, starting with g_1, which is

$$3 \uparrow\uparrow\uparrow\uparrow 3$$

g_1 is a huge number by normal standards – the result of a hexation – but is tiny compared with the moser. After all, the moser is calculated using a (mega – 2) number of up-arrows, whereas g_1 involves just four. But the next step in figuring out Graham's number produces a monstrous jump because g_2 involves g_1 number of up-arrows:

$$g_2 \longrightarrow 3 \overbrace{\uparrow\uparrow \cdots\cdots \uparrow\uparrow}^{} 3$$
$$g_1 \longrightarrow 3 \uparrow\uparrow\uparrow\uparrow 3$$

The moser uses fewer than $10 \uparrow\uparrow 258$ up-arrows and g_2 uses $3 \uparrow\uparrow\uparrow\uparrow 3$ of them, a hexation versus a tetration. It makes little difference that the numbers being operated upon are greater in the case of the moser – the hexation wins by a very wide margin, so that g_2 is much bigger than the moser. We're only at the second step in the calculation of Graham's number and have already left the moser far behind, with still another 63 levels to go. After evaluating each g-number, that many up-arrows are fed into figuring out the next g-number until, our intellects battered into submission, we reach g_{64}. Another way to appreciate the tremendous gulf between the moser and Graham's number is to express each in terms of Conway's chained arrows. The moser is less than $3 \rightarrow 3 \rightarrow 4 \rightarrow 2$, whereas Graham's number lies between $3 \rightarrow 3 \rightarrow 64 \rightarrow 2$ and $3 \rightarrow 3 \rightarrow 65 \rightarrow 2$. The value of that third number in the sequence has an astonishingly powerful effect on the size of the chain as it's unpacked. Graham's number takes over as our new big-number champion by a huge margin – but, again, only temporarily.

The sequence $g_1, g_2, \dots g_n$ is generated very simply by the recurrence relation:

$$g_1 = 3 \uparrow\uparrow\uparrow\uparrow 3, \quad g_n = 3 \uparrow^{g_{n-1}} 3$$

So, it's easy to define a number that's bigger than Graham's number, g_{64}: just put $n = 65$! This takes us to the next step in the g-number sequence, in which we use a Graham's number of up-arrows to produce g_{65}:

$$g_{65} = 3 \uparrow^{\text{Graham's number}} 3$$

Bigger still is g_{66} and much bigger still is g_{googol}. It doesn't take much in the way of creativity to keep naming larger and larger values for n so as to move ever further along the endless line of g-numbers. If you say g_{googol}, I might say

$g_{googolplex}$, to which you might reply $g_{Graham's\ number}$. These are fantastically large numbers, it's true – much bigger than Graham's number. But they're all produced using the same mechanism, or function, as that used to generate Graham's number. They bring nothing new to the table and so, in googological parlance, are mere salad numbers.

Returning to our analogy with space travel, the quest to define bigger and bigger numbers is like the effort to reach more and more remote objects in the universe. In order to send spacecraft to other stars in a reasonable amount of time – say, somewhere between ten and fifty years – we need a propulsion system that can reach much higher speeds than are possible with chemical rockets. Plenty of ideas have been put forward for such systems, including a scheme called Project Daedalus, proposed in the early 1970s, for a large uncrewed vessel, which would use a series of small nuclear fusion explosions to accelerate it to nearby Barnard's Star, six light-years away, in a journey lasting about half a century. To travel to more distant stars, thousands of light-years away, or across intergalactic distances, in sensible amounts of time, something like the warp drive of *Star Trek* would be needed.

Specific objects in space, which lie at greater and greater distances, such as Pluto, Barnard's Star, the supermassive black hole at the heart of the Milky Way, and the Andromeda Galaxy, are analogous to bigger and bigger special numbers – the googol, the mega, Graham's number, and so on. The means of reaching these cosmic objects on practical timescales are, in mathematical terms, equivalent to the increasingly powerful functions needed to generate larger and larger numbers quickly.

In all of this there's a strong human element. In the case of interstellar travel we'd like to get some results back within a few decades at most because that length of time falls within a human lifespan. *Voyagers 1* and 2, already in interstellar space, will eventually reach the distances of the nearest stars but not for tens of thousands of years (by which time the spacecraft will be long dead). If we were immortal, we might be content to wait tens of millennia to see close-up pictures from our robot probes of Alpha Centauri or the Sirius system, but our personal outlook and ambitions are based on a timescale of less than a century.

The same type of thinking transfers to mathematics and our efforts to reach greater and greater distances along the number line. It's commonly assumed that in some Platonic realm all the numbers that can exist do exist, with all their digits laid out in full for eternity, and everything that's mathematically possible has a permanent home. This concept is reminiscent of the 'block universe' of Einstein in which all of space and time, and all the events that have occurred or ever will occur, are portrayed as a complete, unchanging four-dimensional entity. Being human, and therefore with finite abilities and access to resources, we can glimpse only small parts of the Platonic realm of mathematics or the block universe of physics. In the case of the physical cosmos, we use instruments of steadily increasing sensitivity and power, and spacecraft of greater speed and technical sophistication, to bring more distant objects into clearer view. In mathematics, to define and reach bigger and bigger numbers, we turn to faster and faster growing functions.

We know that, in theory, any number, however large, can be reached simply by counting up one at a time, just as we know that it would be possible to get to the edge of the current observable universe by moving at walking speed. But we don't have the luxury of eternity in which to attain our goals in such leisurely ways, nor, in the mathematical case, the physical resources to store the result. We're finite beings with a need for results on a human timescale.

So, in the search for bigger and bigger numbers, we look for ever-faster methods of generating gigantic integers. The benchmark for such functions, as we saw in Chapter 10, is the fast-growing hierarchy (FGH) – a family of recursive functions indexed by ordinal numbers, both finite and transfinite. The key property of this hierarchy is that the output from a function is fed back as the input to the next iteration of the function, and this is repeated for however many times is dictated by the ordinal index.

In mainstream maths and googology, FGH is the gold standard for measuring the strength of other fast-growing functions and the relative sizes of the numbers to which these functions give rise. Often it's not easy, or even possible, to make a direct comparison: using FGH as a benchmark for rival functions isn't as straightforward as measuring, say, distances with a ruler or the passage of time with a stopwatch. We can't pinpoint, for example, exactly where Graham's number lies on the FGH scale. However, it *is* possible to identify the first function in the hierarchy that indisputably *surpasses* g_{64} for a specific input. The reasoning runs like this: f_ω is roughly on a par with the Ackermann function, so that $f_{\omega+1}$ is comparable with the iterated Ackermann function. Graham's number is the same as 64 iterations of the Ackermann number A(4), in other words $A^{64}(4) = A(A(A...A(4)...))$, with 64 parentheses, from which it follows that $f_{\omega+1}(64) > g_{64}$.

The next computable number we looked at in detail, in Chapter 10, was TREE(3). This is the value of the tree function, TREE(n), arising from Kruskal's tree theorem, for $n = 3$. It represents a monumental jump in size from Graham's number. The smallest that TREE(3) could be – its proven lower bound – is more than $A^{A(187,196)-2}(4)$. So, whereas Graham's number is equal to 'only' 64 iterations of $A(4)$, TREE(3) exceeds (and possibly far exceeds) $A(187,196) - 2$ iterations! As to where TREE lies in terms of the fast-growing hierarchy, we know that it grows at least as fast as the FGH at the small Veblen ordinal but it could be at a much higher ordinal.

Just as TREE(3) is incomprehensibly larger than Graham's number (which is itself stunningly big), so the number known as SCG(13) is incomprehensibly larger than TREE(3). Like TREE(n), the subcubic graph function SGC(n) is a superfast-growing function arising from graph theory that was first defined by American logician Harvey Friedman. The specific value SCG(13) has been the subject of much googological research simply because it demonstrably leaves TREE(3) far behind in its wake. In fact, it's been shown that the *minimum* value that SCG(13) could take is $\text{TREE}^{\text{TREE}(3)}(3) = \text{TREE}(\text{TREE}(...\text{TREE}(3)...))$, with TREE(3) sets of parentheses.

Much bigger still than SCG(13) is Loader's number – the output of the winning C program, submitted by Ralph Loader, in the 2001 Bignum Bakeoff competition. Explaining exactly what Loader's number is and proving that it dwarfs even SCG(13) would involve penetrating deeply into aspects of computer science. One of these is the calculus of constructions, which, in turn, is a development of lambda calculus – the model of computing introduced by Alonzo Church in the 1930s. In the calculus of constructions, proofs can be put into the form of binary numbers and expressions written as power towers. The number that would eventually be produced if Loader's program (loader.c) were run on a computer with infinite memory is represented by $D^5(99)$. The function $D(n)$ is a measure of how many expressions in the calculus of constructions can be proved within approximately $\log(n)$ inference steps. The value of $D(n)$ goes up very quickly as n increases because the calculus of constructions is particularly powerful and expressive. David Moews, who organised the Bignum Bakeoff, showed that $D(99)$ is bigger than $2 \uparrow\uparrow 30,419$ and $D^2(99)$, $= D(D(99))$, is much bigger than $f_{\varepsilon 0+\omega^3}(1,000,000)$ on the fast-growing hierarchy. Loader's number is the output of $D^5(99) = D(D(D(D(D(99)))))$ and dwarfs both TREE(3) and SCG(13) to become our new current champion. As to where SCG and Loader's D^5 lie on the fast-growing hierarchy, we simply don't know, except that they far surpass the level of TREE. The lower bound proved by Moews for Loader's number of $f_{\varepsilon 0+\omega^3}(1,000,000)$ is incredibly weak and only chosen since it was enough to show that Loader's number beat the second-largest entry in the Bakeoff contest!

Updating our list of named big numbers, from smallest to largest, we now have:

> **One million, $1,000,000 = 10^6$**
> **Avogadro's number, 6.022×10^{23}**
> **Archimedes' sand number, $\sim 10^{63}$**

Number of fundamental particles in the observable universe, ~10^{86}

The googol, 10^{100}

One centillion, 10^{303}, the largest number ending in –illion named in dictionaries

Largest known prime number, $2^{82,589,933} - 1$ (over 23 million digits long)

The googolplex, $10^{\text{googol}} = 10^{10^{100}}$

Skewes' (first) number, $10^{\wedge}10^{\wedge}10^{\wedge}34$

Poincaré recurrence time for the universe, $10^{\wedge}10^{\wedge}10^{\wedge}10^{\wedge}10^{\wedge}1.1$

The mega, between $10 \uparrow\uparrow 257$ and $10 \uparrow\uparrow 258$

The megiston, between $10 \uparrow\uparrow\uparrow 11$ and $10 \uparrow\uparrow\uparrow 12$

The moser, between $2 \uparrow^{\text{mega-2}} 3$ and $2 \uparrow^{\text{mega-2}} 4$

Graham's number: $3 \to 3 \to 64 \to 2 < g_{64} < 3 \to 3 \to 65 \to 2$; $g_{64} = A^{64}(4)$

TREE(3): $> A^{A(187,296)-2}(4)$

SCG(13): $> \text{TREE}^{\text{TREE}(3)}(3)$

Loader's number, $D^5(99)$

As we move to larger and larger numbers, it gets increasingly difficult to put them in order of size. The main reason for this is that they're defined is such totally different ways. Both the TREE and SCG functions, for instance, arise out of graph theory. Loader's number, on the other hand, is the output of a program, written in fewer than 512 characters in the C language (to conform with the rules of the Bignum Bakeoff), which takes every pattern of bits (1s and 0s) up to some length n and expresses this as a program in the calculus of constructions. It's impossible to know anything specific about Loader's number, such as how many digits it contains or what its final few digits would be if it could be written out in full. However, it *is* possible to analyse the growth of the function at the heart of Loader's number, $D(n)$, and thereby show that after five nested recursions of D with $n = 99$, $D^5(99)$ far exceeds SGC(13). In fact, in all cases, once we get beyond Graham's number territory, we're relying on rough estimates of the growth rates of functions to help us determine the relative sizes of the numbers to which they give rise. We've come across, in this book, a lot of what might seem to be arbitrary numbers, such as TREE(3) and SGC (13). Why not TREE(127) or SGC (42), or any other value that we might pluck out of the air? The answer is that, having established the approximate growth rate for a function, it becomes possible to find the smallest specific input for which the function outputs a value that's demonstrably of an entirely different magnitude to the output of rival functions. In the case of TREE, there's a mind-blowing explosion in size from TREE(2) to TREE(3), which makes TREE(3) an instant member of the big-number hall of fame. SGC(13) then achieved stardom – in googological circles, at any rate! – because it's the smallest of the SGC sequence that can be proved to make TREE(3) look puny by comparison.

Another difficulty in ranking big numbers is that, frankly, relatively few professional mathematicians are interested in the pursuit. It's more of a hobby than a serious area of research. This doesn't mean that some very talented mathematicians aren't involved in googology: if there weren't, no progress would be made at all. In fact, to thoroughly understand what Loader's number, for example, is all about, calls for a graduate-level grasp of computer science and mathematical logic. But it remains true that not much is published in academic journals on the relative sizes of very large numbers. For the most part, any peer review that takes place is confined to online googology forums and the like.

Despite all these challenges, we can confidently state that Loader's number is currently leading the field in our big numbers stakes. It's unquestionably true that:

Loader's number >> Graham's number >> SCG(13) >> TREE(3)

Loader's D^5 isn't the end of the story, though – not by a long way. There's the obvious point that D^6, the next level of recursion, has a higher growth rate than D^5 and that $D^6(99)$, for instance, must inevitably be much bigger than $D^5(99)$. We can become even more tiresome and lacking in originality by arguing that something like $D^{googolplex}$(Graham's number) leaves Loader's original function in the dust. But nobody likes a salad number smart alec. Perhaps to avoid going down such an unprofitable route we should refer to our current big-number champion as 'Loader's number or any higher outputs based on extensions to Loader's function'. The same kind of generalisation could be applied to all the fast-growing functions we've encountered and to any specific large numbers to which they give rise, but from now on we'll take this as given.

Loader's number is certainly one of the biggest *computable* numbers ever defined. But the fact that it's the outcome of a finite, terminating algorithm means that, in theory, it's knowable to any degree of precision. We can't *actually* compute it because there aren't enough resources in the universe. But that's irrelevant. A Turing machine, which, by definition, has access to unlimited amounts of memory and time, could calculate Loader's number, along with any of the other numbers on the list we've put together so far.

As we saw in Chapter 11, however, there are also uncomputable functions, the growth rate of which is guaranteed eventually to exceed that of any computable function. Among the best known of these uncomputable functions are the busy beavers, first described by Tibor Radó in the early sixties. Oddly enough, as we saw, a busy beaver is an incredibly inefficient computer program – in fact, that's the whole point. It's a program that takes the greatest possible number of steps to achieve a given goal. The corresponding busy beaver numbers, BB(n), answer the question: given a certain number of rules, n, what are the most steps that a Turing machine can take before it halts? The sequence of busy beaver numbers climbs at a fantastic rate. BB(2) = 6 and BB(3) = 107, but after that the numbers skyrocket. BB(6) is at least $7.4 \times 10^{36,534}$ and somewhere around BB(17) the busy beaver sequence overtakes Graham's number. Faster and faster the busy beavers accelerate until, inevitably, they overhaul Loader's number and any still larger computable number that we could ever define. No one knows for what value of n, BB(n) overhauls $D^5(99)$. But, for the sake of argument, let's say it's 1,000, so that BB(1,000) > Loader's number. The fact is, whatever number tops our list of computable numbers there'll be some specific busy beaver number that is bound to exceed it.

It's possible to extend the concept of busy beaver numbers by supposing that there are super Turing machines that can solve the halting problem. These higher-level Turing machines would give rise to an even faster-growing sequence of 'super busy beaver numbers', which would in turn be surpassed by super super busy beaver numbers, and so on. But this idea hasn't been well defined and, in any case, is a mere variation on the underlying busy beaver theme.

Hollom's iota function, which by definition would absorb every other function at every moment in time, in the optimal way, so as to generate the largest possible output, is novel but lacks any semblance of formalism. There's no way to analyse the iota function mathematically and it remains little more than an imaginative suggestion and a philosophical exercise.

This leaves just one number truly worthy of adding to the end of our list from the perspective of professional mathematics. It's original in its conception, reasonably well defined, and utterly stupendous in size. By common consent, it's one of the largest – if not the largest – finite numbers ever conceived and, for that reason, we're going to crown it our ultimate champion. The winner of the Big Number Duel and our nominee for the biggest number in the world is: Rayo's number, $R(10^{100})$. Augustin Rayo's formal definition of his eponymous number involves about ten lines of symbols and expressions in the language of set theory, and may as well be written in Chinese for anyone who hasn't gone deeply into the subject at university level. A simplified version of it runs along these lines: 'the smallest positive integer bigger than any finite positive integer named by an expression in the language of first-order set theory with a googol symbols or less'. Making some reasonable assumptions, this boils down to the assertion that Rayo's function, $R(n)$, outgrows all other functions definable under the standard system of axioms (ZFC) used in most of modern mathematics.

223

For now, when it comes to the largest integers that we can sensibly talk about, Rayo's number more or less marks the boundary with the unknown. Attempts have been made to define even bigger numbers, with curious names such as Fish 7, UTTER OBLIVION, and BIG FOOT. The last of these, announced in 2014 by a googologist with the monicker Wojowu, uses a strategy similar to that used to construct Rayo's number, but applied to a new domain that Wojowu called the oodleverse. To penetrate this weird world would involve first learning the language of first-order oodle theory (an extension of standard first-order set theory) – a venture best tackled with a higher degree in maths and a wry sense of humour. Unfortunately, it transpires that oodle theory has inconsistencies, which render BIG FOOT ill-defined.

Both Loader's number and Rayo's number, which are at or near the top of our league tables for the largest countable and uncountable numbers, respectively, came about because of competitions. Maybe it's time for another contest to spark some brand new ideas! Until then, in their efforts to trek ever further down the number line, googologists may need to rely on the same tricks used by Rayo but apply them to beefed-up versions of first-order set theory. Cosmologists strive to see more clearly to the edge of the observable universe and find ways to learn about what may lie beyond; mathematicians look to the farthest limits of their own subject.

Meanwhile, we can't ignore any longer the elephant that's been sitting in the room since the start of this chapter. We know full well that the smallest of transfinite numbers, aleph-null, is bigger than any finite number we could ever possibly name. We know, too, that there are infinitely many alephs, each infinitely bigger than the one before. Yet mathematicians can dream of cardinals whose size exceeds that of any conceivable aleph. To do this, they have to move beyond the usual foundations of their subject and resort to what are called forcing axioms – a technique pioneered by Stephen Kleene, mentioned earlier. This leads to the concept of the modestly named 'large cardinals', which in reality are spectacularly vast, including those with special names such as Mahlo cardinals and supercompact cardinals.

Finally (at least for now), there's the notion of Absolute Infinity, sometimes represented by capital omega (Ω) – an infinity that transcends or surpasses all others. Cantor himself spoke about it but mainly in religious terms. He was a deeply committed Lutheran whose Christian convictions occasionally surfaced in his scholarly work. To him, Omega, if it existed, could only do so in the mind of the god in which he believed. On that basis, Omega is nothing more than a grand metaphysical speculation. Sticking purely to mathematics, Absolute Infinity can't be defined rigorously, so mathematicians, unless they allow philosophical speculation to get the better of them, tend to ignore it. There may be the temptation to characterise it as the number of elements in the universe of all sets – the so-called von Neumann universe. But the von Neumann universe isn't actually a set (rather, it's a class of sets), so that it can't be used to define any specific kind of infinity, whether cardinal or ordinal. More controversially, Omega might be thought of as the most sensible result if 1 were to be divided by 0. This isn't a procedure normally defined in maths, though it can be done in certain forms of geometry, for example, projective geometry, which yield the idea of a 'point at infinity' or a 'line at infinity'. The quest for Omega will continue to challenge future generations of mathematicians, logicians and philosophers. Meanwhile, we have plenty of infinities, each infinitely larger than the one before, to keep our brains occupied.

What is the biggest number in the world? It depends on which 'world' we're talking about. In that part of the mathematical cosmos we've so far explored we could sensibly answer 'Rayo's number'. But the future will hold many surprises and our journey into the limitless depths of the numerical cosmos has barely begun.

Acknowledgements

DAVID: My thanks go as always to my family, especially my wife Jill for her tireless love and encouragement. Thank you also to my mathematical friend Andrew Barker for his many helpful comments and suggestions.

AGNIJO: I'd like to thank my parents for everything they do for me, and my brother Aaryan for inspiring me every day.

We're both grateful to our editor at Oneworld, Sam Carter, for his expert advice and guidance throughout the writing of this book and the earlier *Weird Maths* series. Our thanks go, too, to all the other wonderful people at Oneworld who have helped bring this project to fruition.

Bibliography

Aczel, A. D. *The Mystery of the Aleph: Mathematics, the Kabbalah, and the Search for Infinity*. (New York: Pocket Books, 2001)

Asimov, I. *Skewered!, Of Matters Great and Small*. (New York: Ace Books, 1976)

Boolos, G. S., Burgess, J. P. and Jeffrey, R. C. *Computability and Logic* (5th ed.). (Cambridge: Cambridge University Press, 2007)

Conway, J. H. and Guy, R. K. *The Book of Numbers*. (New York: Springer-Verlag, 1996)

Darling, D. and Banerjee, A. *Weird Maths*. (London: Oneworld, 2018)

Davis, P. J. *The Lore of Large Numbers*. (New York: Random House, 1961)

Drake, F. R. *Set Theory: An Introduction to Large Cardinals*. (New York: Elsevier Science, 1974)

Elliott, A. *Is That a Big Number?*. (Oxford: Oxford University Press, 2018)

Gamow, G. *One, Two, Three... Infinity: Facts and Speculations of Science*. (London: Viking, 1947; reprinted in paperback by Dover, 1988)

Kanigel, R. *The Man Who Knew Infinity: A Life of the Genius Ramanujan*. (New York: Washington Square Press, 1991)

Kasner, E. and Newman, J. *Mathematics and the Imagination*. (New York: Simon and Schuster, 1940)

Lavine, S. *Understanding the Infinite*. (Cambridge, Mass.: Harvard University Press, 1994)

Moore, A. W. *The Infinite*. (New York: Routledge, 1990)

Nowlan, R. A. 'Large and Small', in *Masters of Mathematics*. (Rotterdam: Sense Publishers, 2017), pp. 217–27

Rucker, R. *Infinity and the Mind*. (Princeton, New Jersey: Princeton University Press, 2005)

Schwartz, R. E. *Really Big Numbers*. (Providence, Rhode Island: American Mathematical Society, 2014)

Wallace, D. F. *Everything and More: A Compact History of Infinity*. (New York and London: W.W. Norton & Company, Inc., 2004)

Wells, D. *The Penguin Dictionary of Curious and Interesting Numbers*. (London: Penguin Books, 1997)

Useful websites and webpages

BAEZ, JOHN. AZIMUTH BLOG. *ENORMOUS INTEGERS* PAGE.
https://johncarlosbaez.wordpress.com/2012/04/24/enormous-integers/

COOKIE FONSTER'S LARGE NUMBER SITE.
https://sites.google.com/site/pointlesslargenumberstuff/home/l/timeline

GIROUX STUDIO'S SERIES ON EXTREMELY LARGE NUMBERS.
https://www.youtube.com/watch?v=vq2BxAJZ4Tc&list=PLUZ0A4xAf7nkaYHtnqVDbHnrXzVAOxYYC&index=1

GOOGOLOGY WIKI ON THE FAST-GROWING HIERARCHY.
http://googology.wikia.com/wiki/Fast-growing_hierarchy

GOUCHER, ADAM. ARTICLES ON FAST GROWING FUNCTIONS.
https://cp4space.hatsya.com/2013/01/06/fast-growing-4/

HOLLOM, LAWRENCE. BIG NUMBER PAGES.
https://sites.google.com/a/hollom.com/extremely-big-numbers/old-homepage

METZLER, DAVID. YOUTUBE SERIES: 'RIDICULOUSLY HUGE NUMBERS'.

https://www.youtube.com/watch?v=QXliQvd1vW0

MUNAFO, ROBERT P. 'LARGE NUMBERS' PAGES.

http://www.mrob.com/pub/math/largenum.html

NUMBERPHILE. HOW BIG IS GRAHAM'S NUMBER? (FEATURING RON GRAHAM).

https://www.youtube.com/watch?v=GuigptwlVHo

NUMBERPHILE. RON GRAHAM AND GRAHAM'S NUMBER.

https://www.youtube.com/watch?v=rGWuimr5CGQ

SAIBIAN, SBIIS. LARGE NUMBER SITE.

https://sites.google.com/site/largenumbers/

STEPNEY, SUSAN. UNIVERSITY OF YORK. BIG NUMBER PAGE.

https://www-users.cs.york.ac.uk/~susan/cyc/b/big.htm

References

CHAPTER 1: OF SAND AND STARS

Ifrah, G. *The Universal History of Numbers*. (London: Harvill, 2000)

Knuth, D. E. 'Supernatural Numbers', in *The Mathematical Gardner*, ed. D. Klarner. (Boston, Mass.: Springer, 1981), pp. 310–25

Nowlan, R. A. 'Large and Small', in *Masters of Mathematics*. (Rotterdam: Sense Publishers, 2017), pp. 217–27

Vardi, I. *Archimedes, The Sand Reckoner*. http://www.lix. polytechnique.fr/Labo/Ilan.Vardi/sand_reckoner.ps

CHAPTER 2: AT THE LIMITS OF REALITY

Bekenstein, J. D. (1981). 'Universal upper bound on the entropy-to-energy ratio for bounded systems', *Physical Review* D. 23 (2), pp. 287–98

Dirac, P. A. M. (1974). 'Cosmological models and the Large Numbers hypothesis', *Proceedings of the Royal Society* of London. A. *Mathematical and Physical Sciences*, 338 (1615), pp. 439–46

Eddington, A. *The Mathematical Theory of Relativity*. (Cambridge: Cambridge University Press, 1923)

Lloyd, S. (2000). 'Ultimate physical limits to computation', *Nature*, 406 (6799), pp. 1047–54

Markov, I. (2014). 'Limits on Fundamental Limits to Computation', *Nature*, 512 (7513), pp. 147–54

Page, D. N. (1994). 'Information Loss in Black Holes and/or Conscious Beings?'. https://arxiv.org/pdf/hep-th/9411193. pdf

CHAPTER 3: MATHS UNBOUND

Anderson J. (2004). 'Iterated exponentials', *The American Mathematical Monthly*, 111 (8), pp. 668–79

Bloch, W. G. *The Unimaginable Mathematics of Borges' Library of Babel*. (Oxford: Oxford University Press, 2008)

Moroni, L. (2019). 'The strange properties of the infinite power tower'. https://arxiv.org/pdf/1908.05559.pdf

Shannon, C. (1950). 'XXII. Programming a Computer for Playing Chess', *Philosophical Magazine*, series 7, 41 (314)

Skewes, S. (1933). 'On the Difference $\pi(x) - \mathrm{Li}(x)$', *Journal of the London Mathematical Society*, 8, pp. 227–83

Skewes, S. (1955). 'On the Difference $\pi(x) - \mathrm{Li}(x)$, II', *Proceedings of the London Mathematical Society*, 5, pp. 48–70

REFERENCES

CHAPTER 4: UP, UP AND AWAY

Guy, R. K. and Selfridge, J. L. (1973). 'The Nesting and Roosting Habits of the Laddered Parenthesis', *American Mathematical Monthly*, 80, pp. 868–76

Knuth, D. E. (1976). 'Mathematics and Computer Science: Coping with Finiteness', *Science*, 194 (4271), pp. 1235–42

Steinhaus, H. *Mathematical Snapshots*. (Oxford: Oxford University Press, 1950)

CHAPTER 5: G WHIZZ

Barkley, J. (2008). 'Improved lower bound on a Euclidean Ramsey problem'. https://arxiv.org/pdf/0811.1055.pdf

Gardner, M. (1977). 'Mathematical Games', *Scientific American*, 237 (5), pp. 18–28

Graham, R. L. and Rothschild, B. L. (1971). 'Ramsey's theorem for n-parameter sets', *Transactions of the American Mathematical Society*, 159, pp. 257–92

Graham, R. L. and Rothschild, B. L. 'Ramsey Theory', in *Studies in Combinatorics*, ed. G-C Rota. (Washington: Mathematical Association of America, 1978), pp. 80–99

Lavrov, M., Lee, M. and Mackey, J. (2014). 'Improved upper and lower bounds on a geometric Ramsey problem', *European Journal of Combinatorics*, 42, pp. 135–44

McWhirter, N. *Guinness Book of World Records*. (New York: Sterling, 1980), p. 193

CHAPTER 6: CONWAY'S CHAINS

Bailer-Jones, C. A. and Farnocchia, D. (2019). 'Future Stellar Flybys of the *Voyager* and *Pioneer* Spacecraft', *Research Notes of the American Astronomical Society*, 3 (4)

Conway, J. H. and Guy, R. K. *The Book of Numbers*. (New York: Springer-Verlag, 1996)

Gardner, M. (1970). 'Mathematical Games – The fantastic combinations of John Conway's new solitaire game "life"', *Scientific American*, 223 (4), pp. 120–3

Roberts, S. *Genius at Play: The Curious Mind of John Horton Conway*. (New York: Bloomsbury, 2015)

CHAPTER 7: ACKERMANN AND THE POWER OF RECURSION

Ackermann, W. (1928). 'Zum Hilbertschen Aufbau der reellen Zahlen', *Mathematische Annalen*, 99, pp. 118–33

Dötzel, G. (1991). 'A function to end all functions', *Algorithm: Recreational Programming*, 2, pp. 16–17

Reid, C. *Hilbert*. (New York: Copernicus, 2012)

CHAPTER 8: FIGURE THIS – IF YOU CAN

Gardner, M. (1977). 'Mathematical Games', *Scientific American*, 237 (5), pp. 18–28

Hodel, R. E. *An Introduction to Mathematical Logic*. (New York: Dover, 2013)

Minsky, M. *Computation: Finite and Infinite Machines*. (Englewood Cliffs, New Jersey: Prentice Hall, 1967)

Smith, P. *An Introduction to Gödel's Theorems*. (Cambridge: Cambridge University Press, 1967)

Sudkamp, T. A. *Languages and Machines: An Introduction to the Theory of Computer Science*. (New York: Pearson, 2005)

Turing, A. M. (1937). 'On Computable Numbers, with an Application to the *Entscheidungsproblem*', *Proceedings of the London Mathematical Society*, s2–42, pp. 230–65

CHAPTER 9: INFINITE MATTERS

Cantor, G. *Contributions to the Founding of the Theory of Transfinite Numbers*, ed. P. Jourdain. (New York: Dover, 1955)

Dauben, J. W. (1983). 'Georg Cantor and the Origins of Transfinite Set Theory', *Scientific American*, 248 (6), pp. 122–31

CHAPTER 10: GROWING FAST

Buchholz, W. and Wainer, S. S. 'Provably Computable Functions and the Fast Growing Hierarchy', in *Logic and Combinatorics*, ed. S. Simpson, vol. 65. (Providence, Rhode Island: American Mathematical Society, 1987), pp. 179–98

Schmidt, D. (1977). 'Built-up systems of fundamental sequences and hierarchies of number-theoretic functions', *Archiv für mathematische Logik und Grundlagenforschung*, 18, pp. 47–53

Weiermann, A. (1997). 'Sometimes slow growing is fast growing', *Annals of Pure and Applied Logic*, 90 (1–3), pp. 91–9

CHAPTER 11: DOES NOT COMPUTE!

Aaronson, S. (2020). 'The Busy Beaver Frontier', *ACM SIGACT News*, 51 (3), pp. 32–54

Ben-Amram, A. M. and Petersen, H. (2002). 'Improved Bounds for Functions Related to Busy Beavers', *Theory of Computing Systems*, 35, pp. 1–11

Chaitin, G. J. 'Computing the Busy Beaver Function', in *Open Problems in Communication and Computation*, ed. T. M. Cover and B. Gopinath. (New York: Springer, 1987), pp. 108–12

Dewdney, A. K. (1984). 'Computer Recreations: A computer trap for the busy beaver, the hardest working Turing machine', *Scientific American*, 251 (2), pp. 10–17

Dewdney, A. K. (1985). 'Five-state Busy Beaver Turing Machine Contender', *Scientific American*, 252 (4), p. 30

Marxen, H. and Buntrock, J. (1990). 'Attacking the Busy Beaver 5', *Bulletin of the EATCS*, 40, pp. 247–51

Michel, P. (2009). 'The busy beaver competition: a historical survey'. https://arxiv.org/pdf/0906.3749.pdf

Radó, T. (1962). 'On non-computable functions', *Bell System Technical Journal*, 41 (3), pp. 877–84

Siegelmann, H. T. (1995). 'Computation Beyond the Turing Limit', *Science*, 268 (5210), pp. 545–8

CHAPTER 12: THE STRANGE WORLD OF THE GOOGOLOGIST

Aaronson, S. (1999). 'Who Can Name the Bigger Number?' https://www.scottaaronson.com/writings/bignumbers.html

Aaronson, S. (2020). 'The Busy Beaver Frontier', *ACM SIGACT News*, 51 (3), pp. 32–54

Michel, P. (1993). 'Busy beaver competition and Collatz-like problems', *Archive for Mathematical Logic*, 32 (5), pp. 351–67

CHAPTER 13: BRIDGE TO BEYOND

Eriksson, K., Estep, D. and Johnson, C. '17 Do Mathematicians Quarrel? §17.7 Cantor Versus Kronecker', in *Applied Mathematics: Body and Soul: Volume 1: Derivatives and Geometry in IR3*. (New York: Springer, 2003), pp. 230–2

Hollom, L. *Iota function*. https://sites.google.com/a/hollom.com/extremely-big-numbers/old-homepage/iota-function (accessed 15 November 2020)

Patey, L. and Yokoyama, K. (2018). 'The proof-theoretic strength of Ramsey's theorem for pairs and two colors', *Advances in Mathematics*, 330, pp. 1034–70

Tait, W. W. 'Remarks on finitism', in *Reflections on the Foundations of Mathematics: Essays in Honor of Solomon Feferman*, ed. W. Sieg, R. Sommer and C. Talcott. (Natick, Mass.: A K Peters Ltd., 2002), pp. 407–16

Tiles, M. *The Philosophy of Set Theory: An Historical Introduction to Cantor's Paradise*. (New York: Dover, 2004)

Wolchover, N. (2013). 'Dispute over Infinity Divides Mathematicians', *Scientific American*. https://www.scientificamerican.com/article/infinity-logic-law/

CHAPTER 14: THE BIGGEST NUMBER OF ALL

Crandall, R. E. (1997). 'The Challenge of Large Numbers', *Scientific American*, 276 (2), pp. 74–8

Singh, D. and Singh, J. N. (2007). 'von Neumann Universe: A Perspective', *International Journal of Contemporary Mathematical Sciences*, 2, pp. 475–8

ABOUT THE AUTHORS

David has a PhD in astronomy from the University of Manchester. For the past thirty-five years he's been a freelance science writer and is the author of around fifty books on subjects such as cosmology, physics, philosophy and mathematics. His website, www.daviddarling.info, and YouTube channel, discovermaths, are widely used online resources. He is currently producing a science-themed musical show called 'The Science Fiction Experience'.

Agnijo was born in Kolkata, India, but has spent most of his life in Scotland. His extraordinary mathematical talents were recognised at an early age. In 2018 he came joint first in the International Mathematical Olympiad, recording a perfect score and affirming his status as one of the world's most outstanding young mathematicians. He is now continuing his studies at Trinity College, Cambridge.

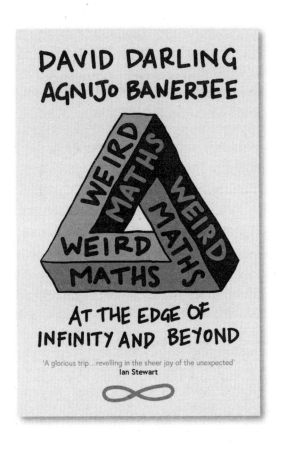

DAVID DARLING
AGNIJO BANERJEE

WEIRD MATHS
MATHS
WEIRD MATHS
WEIRD MATHS

AT THE EDGE OF
INFINITY AND BEYOND

'A glorious trip...revelling in the sheer joy of the unexpected'
Ian Stewart

David Darling and Agnijo Banerjee draw connections
between the cutting edge of modern maths and life itself,
on a quest to consider the existence of free will, how to
see in 4D and the future of quantum computers. Packed
with puzzles and paradoxes, this is for anyone who wants
life's questions answered – even those you never thought
to ask.